PRES DA

Volume 1 - Diesel & Electric

2nd Edition

Iain Dobson &
Neil Webster

ISBN 0-947773-50-9

© Copyright 1994. Metro Enterprises Ltd., 312 Leeds Road, Birstall, Batley, WF17 0HS.

NOTES

PRESERVATION DATAFILE is produced as a budget priced handy pocket size guide to former BR locomotives & multiple units now in preservation or industrial use. This volume covers Diesel & Electric Vehicles, whilst Volume 2 covers Steam Vehicles.

INDEX

LAYOUT OF INFORMATION

All Sections

Technical data relates to as built condition and takes no account of any subsequent modifications.

Section 1.1

Locomotives are arranged in class number order. The tabular details are arranged in columns as shown in the following example:

55015   D9015   01.82   Midland Railway Centre   OP

Column 1 is the TOPS number. Where more than one number has been carried over the years by a particular locomotive, the number carried carried for the greatest length of time is shown, with details of other numbers carried given as notes.

Column 2 is the 1957 series number (where applicable).

Column 3 is the withdrawal date from BR service (excluding periods in use for other than traction purposes.

Column 4 is the current location (or intended location of recently purchased vehicles).

Column 5 is the current status (see status code list below).

Sections 1.2-1.5

Locomotives are arranged in number order, 1948 series numbers preceding 1957 numbers and Departmental numbers. The tabular details are arranged in columns as shown in the following example:

18000	02.60	The Railway Age, Crewe	IE

Column 1 is the locomotive number. Where more than one number has been carried by a particular locomotive, other numbers are given as notes.

Column 2 is the withdrawal date from BR or WD service.

Column 3 is the current location.

Column 4 is the current status (see status code list below).

Sections 2.1-2.4

Vehicles are arranged in numerical order of final number carried on BR. The tabular details are arranged in columns as shown in the following example:

53528	50528	03.92	Llangollen Railway	OP

Column 1 is the final number carried on BR.

Column 2 is the previous number carried on BR (where applicable).

Column 3 is the withdrawal date from BR service.

Column 4 is the current location (or intended location of recently purchased vehicles).

Column 5 is the current status (see status code list below).

Sections 3.1-3.5

Vehicles are arranged in order of classes, then order of vehicle numbers. The tabular details are arranged in columns as shown in the following example:

29289	DTSO	03.85	Wirral MDC at Kirkdale EMU Depot	SU

Column 1 is the vehicle number.

Column 2 is the type of vehicle within a set (as applicable) or (in section 3.5) the vehicle name.

Column 3 is the withdrawal date from BR service.

Column 4 is the current location (or intended location of recently purchased vehicles).

Column 5 is the current status (see status code list below).

STATUS CODES

CC	In use as a Camping Coach.
CE	Complete Museum Exhibit.
EX	Exported, current fate unknown.
IE	Incomplete Museum Exhibit.
ML	Approved for main line running on British Rail.
OP	Operational, but not approved for main line running.
OH	Operational as hauled stock only (multiple unit vehicles).
SB	Stored on BR.
SP	Retained for use as spares only.
SS	Stored in a serviceable condition.
SU	Stored in an unserviceable condition.
UR	Undergoing repairs.

Where the status column in blank, the current status of the vehicle is not known.

OTHER ABBREVIATIONS

Other abbreviations used in this book are as follows:

ac	Alternating current.
BE	Battery Electric.
BR	British Railways.
dc	Direct current.
DE	Diesel-electric.
DH	Diesel-hydraulic.
DM	Diesel-mechanical.
DMBC	Driving Motor Brake Composite.
DMBS	Driving Motor Brake Second.
DMBSL	Driving Motor Brake Second with lavatory.
DMBT	Driving Motor Brake Third.
DMCL	Driving Motor Composite with lavatory.
DMPMV	Driving Motor Parcels & Miscellaneous Van.
DMS	Driving Motor Second.
DML	Driving Motor Second with lavatory.
DTBS	Driving Trailer Brake Second.
DTCL	Driving Trailer Composite with lavatory.
DTS	Driving Trailer Second.
ED	Third Rail Electric/Diesel-electric.
GTE	Gas Turbine-electric.

GWR	Great Western Railway.
Hz	Hertz.
kV	Kilovolts.
kW	Kilowatts.
LMS	London Midland & Scottish Railway.
LNER	London & North Eastern Railway.
M	Motor coach.
m.	Metres.
mm.	Millimetres.
mph	Miles per hour.
OE	Overhead Electric.
PC	Power Car.
RE	Third Rail Electric.
rpm	Revolutions per minute.
SR	Southern Railway.
t	tons.
TBF	Trailer Brake First with lavatory.
TBFKL	Trailer Brake First (Corridor) with lavatory.
TBS	Trailer Brake Second.
TBSL	Trailer Brake Second with lavatory.
TC	Trailer Composite. (Trailer Car - APT-E only).
TCL	Trailer Composite with lavatory.
TFK	Trailer First (Corridor).
TFLK	Trailer First (Corridor) with lavatory.
TS	Trailer Second.
TSK	Trailer Second (Corridor).
TSL	Trailer Second with lavatory.
TSLRB	Trailer Second Miniature Buffet with lavatory.
v.	Volts.

COVER PHOTOGRAPHS

Front: An immaculate E3003 on display at Crewe Basford Hall on 21.07.94

(Richard Lillie)

Back: 27001 on the Bo'ness and Kinneil line in April 1993.   (Dave Campbell)

PART ONE - LOCOMOTIVES

1.1 LOCOMOTIVES CLASSIFIED UNDER THE TOPS SYSTEM

CLASS 01   0-4-0 DM

Built: 1956 by Andrew Barclay, Kilmarnock.
Engine: Gardner 6L3 of 114 kW at 1200 rpm.

Length over buffers: 7.214 m.	**Weight:** 25.05 t.
Extreme width: 2.578 m.	**Wheel Diameter:** 965 mm.
Extreme height: 3.616 m.	**Maximum Permitted Speed:** 14.25 mph

Original Numbers: 11503 & 11506 respectively.

D2953	06.66	South Yorkshire Railway	OP
D2956	05.66	East Lancashire Railway	UR

CLASS 02   0-4-0 DH

Built: 1960-61 by Yorkshire Engine Company, Sheffield.
Engine: Rolls Royce C6NFL Series 176 of 126 kW at 1800 rpm.

Length over buffers: 6.702 m.	**Weight:** 28.15 t.
Extreme width: 2.591 m.	**Wheel Diameter:** 1067 mm.
Extreme height: 3.486 m.	**Maximum Permitted Speed:** 19.5 mph.

02003	D2853	06.75	Anglo-Charrington, Pensnett	SU
	D2854	02.70	South Yorkshire Railway	OP
	D2858	02.70	Butterley Engineering Co, Ripley	OP
	D2860	12.70	National Railway Museum	OP
	D2866	02.70	Caledonian Railway	SU
	D2867	09.70	Redland Roadstone, Barrow on Soar	OP
	D2868	12.69	Anglo-Charrington, Pensnett	SU

CLASS 03   0-6-0 DM

Built: 1958-62 by BR at Swindon or Doncaster Works.
Engine: Gardner 8L3 of 152 kW at 1200 rpm.

Length over buffers: 7.925 m.	**Weight:** 30.20 t.
Extreme width: 2.591 m.	**Wheel Diameter:** 1092 mm.
Extreme height: 3.720 m.	**Maximum Permitted Speed:** 28.5 mph.

Original Numbers: 03371 was originally Departmental No. 92.
Note: † expected to move to the South Yorkshire Railway during 1995.

03018	D2018	11.75	Mayer Parry Recycling, Willesden	SU
	D2019	07.71	ISA, Ospitaletto, Italy	OP
03020	D2020	12.75	Mayer Parry Recycling, Snailwell	SU
03022	D2022	11.82	Swindon & Cricklade Railway	OP
	D2023	07.71	Kent & East Sussex Steam Railway	OP
	D2024	07.71	Kent & East Sussex Steam Railway	SU
03027	D2027	01.76	Private Site at Darley Abbey	SU
	D2032	08.71	ISA, Ospitaletto, Italy	OP
	D2033	12.71	Siderurgica, Montirone, Italy	SU
	D2036	12.71	Siderurgica, Montirone, Italy	OP
03037	D2037	09.76	BCOE, Oxcroft Disposal Point	OP
	D2041	02.70	Colne Valley Railway	OP
	D2046	10.71	Gulf Oil, Waterston	SU
	D2051	10.71	Ford Motor Company, Dagenham	OP
03059	D2059	07.87	Isle of Wight Steam Railway	OP
03062	D2062	12.80	Dean Forest Railway	UR
03063	D2063	11.87	Colne Valley Railway	UR
03066	D2066	01.88	South Yorkshire Railway	OP
03069	D2069	12.83	Gloucestershire Warwickshire Railway	OP
	D2070	11.71	Cheddleton Railway Centre	OP
03072	D2072	03.81	Lakeside & Haverthwaite Railway	OP
03073	D2073	03.89	The Railway Age, Crewe	OP
03078	D2078	01.88	North Tyneside Steam Railway	OP
03081	D2081	12.80	Genappe Sugar Factory, Charleroi, Belgium	EX
03084	D2084	07.87	Private Site at Darley Abbey	UR
03089	D2089	11.87	Mangapps Farm Railway Museum	OP
03090	D2090	07.76	National Railway Museum	OP
03094	D2094	01.88	South Yorkshire Railway	OP
03099	D2099	02.76	South Yorkshire Railway	UR
03112	D2112	07.87	Nene Valley Railway	OP
03113	D2113	08.75	Heritage & Maritime Museum, Milford Haven	IE
	D2117	10.71	Lakeside & Haverthwaite Railway	OP
	D2118	06.72	Costain Dow-Mac, Tallington	SS
03119	D2119	02.86	Dean Forest Railway	OP
03120	D2120	02.86	Fawley Hill Railway	OP
03128	D2128	07.76	Peak Rail	OP
	D2133	07.69	Courtaulds, Bridgwater	OP
03134	D2134	07.76	Stoomcentrum, Maldegem, Belgium †	UR
	D2138	05.69	Midland Railway Centre	OP
	D2139	05.68	South Yorkshire Railway	OP
03141	D2141	07.85	J.M. DeMulder, Shilton	SU
03144	D2144	02.86	MOD Bicester Military Railway	SU

03145	D2145	07.85	J.M. DeMulder, Shilton	SU
	D2148	11.72	Southport Railway Centre	OP
	D2150	11.72	British Salt, Middlewich	OP
03152	D2152	10.83	Swindon & Cricklade Railway	OP
03158	D2158	07.87	Private site at Darley Abbey	UR
03162	D2162	03.89	Wirral MDC at the Llangollen Railway	OP
03170	D2170	03.89	Otis Euro-Transrail, Salford	OP
	D2178	09.69	Caerphilly Railway Centre	OP
03180	D2180	03.84	South Yorkshire Railway	SU
	D2182	05.68	Gloucestershire Warwickshire Railway	UR
	D2184	12.68	Colne Valley Railway	OP
03189	D2189	03.86	Southport Railway Centre	UR
	D2192	01.69	Paignton & Dartmouth Steam Railway	OP
03196	D2196	06.83	Steamtown Railway Centre	OP
03197	D2197	07.87	South Yorkshire Railway	UR
	D2199	06.72	South Yorkshire Railway	OP
03371	D2371	11.87	Rowden Mill Station Museum	OP
	D2381	06.72	Steamtown Railway Centre	OP
03399	D2399	07.87	Mangapps Farm Railway Museum	OP

CLASS 04   0-6-0 DM

Built: 1952-61 by English Electric, Vulcan Foundry, Newton-le-Willows or Robert Stephenson & Hawthorn, Darlington.
Engine: Gardner 8L3 of 152 kW at 1200 rpm.
Length over buffers: 7.931 m.   **Weight:** 30.25 t.
Extreme width: 2.591 m.   **Wheel Diameter:**1092 (* 991,†1067) mm
Extreme height: 3.693 m.
Maximum Permitted Speed: 27.8 (* 25.8) mph.
Original Numbers: 11103 (D2203), 11106 (D2205), 11108 (D2207), 11122 (D2216), 11135 (D2229), 11151 (D2232), 11215 (D2245), 11216 (D2246).

	D2203*	12.67	Embsay Steam Railway	OP
	D2205*	07.69	Avon & Radstock Railway	SU
	D2207*	12.67	North Yorkshire Moors Railway	OP
	D2216†	05.71	ISA, Ospitaletto, Italy	OP
	D2229†	12.69	South Yorkshire Railway	OP
	D2232†	03.68	Rome, Italy	SU
	D2245†	12.68	Battlefield Steam Railway	OP
	D2246†	07.68	South Yorkshire Railway	OP
	D2267	12.69	Ford Motor Company, Dagenham	OP
	D2271	10.69	West Somerset Railway	OP
	D2272	10.70	British Fuels, Blackburn	OP
	D2279	05.71	East Anglian Railway Museum	OP
	D2280	03.71	Ford Motor Company, Dagenham	SP

D2284	04.71	South Yorkshire Railway	SU
D2295	04.71	Acciaierie, Lonato, Italy	OP
D2298	12.68	Buckinghamshire Railway Centre	OP
D2302	06.69	South Yorkshire Railway	OP
D2310	01.69	South Yorkshire Railway	OP
D2324	07.68	Redland Roadstone, Barrow on Soar	OP
D2325	07.68	Mangapps Farm Railway Museum	OP
D2334	07.68	South Yorkshire Railway	OP
D2337	07.68	South Yorkshire Railway	SU

CLASS 05   0-6-0 DM

Built: 1956-60 by Hunslet Engine Company, Leeds.
Engine: Gardner 8L3 of 152 kW at 1200 rpm.
Length over buffers: 7.722 m.   **Weight:** 30.90 t.
Extreme width: 2.515 m.   **Wheel Diameter:** 1143 (* 1016) mm.
Extreme height: 3.353 m.   **Maximum Permitted Speed:** 17.8 mph.
Original Number: D2554 was originally 11140.
Subsequent Number: 05001 later became 97803 in Departmental Service.

05001†	D2554*	09.83	Isle of Wight Steam Railway	OP
	D2578	06.67	H.P. Bulmer, Hereford	SS
	D2587	12.67	East Lancashire Railway	OP
	D2595	06.68	Southport Railway Centre	OP

CLASS 06   0-4-0 DM

Built: 1959 by Andrew Barclay, Kilmarnock.
Engine: Gardner 8L3 of 152 kW at 1200 rpm.
Length over buffers: 7.899 m.   **Weight:** 36.75 t.
Extreme width: 2.565 m.   **Wheel Diameter:** 1092 mm.
Extreme height: 3.612 m.   **Maximum Permitted Speed:** 22.8 mph.
Subsequent Number: 97804 in Departmental Service.

06003	D2420	09.84	South Yorkshire Railway	OP

CLASS 07   0-6-0 DE

Built: 1962 by Ruston & Hornsby, Lincoln.
Engine: Paxman 6RPHL Mk. III of 204 kW at 1360 rpm.
Length over buffers: 8.166 m.   **Weight:** 42.25 t.
Extreme width: 2.591 m.   **Wheel Diameter:** 1067 mm.
Extreme height: 3.912 m.   **Maximum Permitted Speed:** 20 mph.

07001	D2985	07.77	South Yorkshire Railway	OP
07005	D2989	07.77	ICI, Wilton	SU

	D2991	05.73	BRML, Eastleigh	OP
07010	D2994	10.76	Avon Valley Railway	OP
07011	D2995	07.77	ICI, Wilton	OP
07012	D2996	07.77	South Yorkshire Railway	SU
07013	D2997	07.77	South Yorkshire Railway	SU

CLASS 08   0-6-0 DE

Built: 1952-61 by BR at Derby Locomotive, Darlington, Crewe or Horwich Works.
Engine: English Electric 6KT of 297 kW at 680 rpm.
Length over buffers: 8.915 m.   **Weight:** 49.00 (* 48.60) t.
Extreme width: 2.915 m.   **Wheel Diameter:** 1372 mm.
Extreme height: 3.877 m.   **Maximum Permitted Speed:** 20 mph.
Original Numbers: D3000-D3336/66 were originally 13000-13336/66 in sequence.

	D3000	07.72	South Yorkshire Railway	UR
	D3002	07.72	Plym Valley Railway	OP
	D3014	10.72	Paignton & Dartmouth Steam Railway	OP
08011	D3018	12.91	Chinnor & Princes Risborough Railway	OP
	D3019	06.73	South Yorkshire Railway	SU
08015	D3022	09.80	Severn Valley Railway	OP
08016	D3023	05.80	South Yorkshire Railway	OP
08021	D3029	04.86	Birmingham Railway Museum	OP
08022	D3030	03.86	Arthur Guinness, Park Royal	OP
08032	D3044	08.74	Foster Yeoman, Merehead	
	D3047	07.73	LAMCO Railroad, Liberia	SU
08046	D3059	05.80	Caledonian Railway	OP
08054	D3067	12.80	TILCON, Grassington	OP
08060	D3074	06.84	Arthur Guinness, Park Royal	OP
08064	D3079	12.84	National Railway Museum	UR
	D3092*	10.72	LAMCO Railroad, Liberia	SU
	D3094*	10.72	LAMCO Railroad, Liberia	SU
	D3098*	10.72	LAMCO Railroad, Liberia	SU
	D3100*	10.72	LAMCO Railroad, Liberia	SU
	D3101*	05.72	Great Central Railway	UR
08077	D3102	11.77	RFS(E) Ltd. (Hire Locomotive 007)	OP
08102	D3167	03.88	Grimsby & Louth Railway	SU
08108	D3174	07.84	Kent & East Sussex Steam Railway	OP
08113	D3179	03.84	PD Fuels, Gwaun-Cae-Gurwen DP	OP
08114	D3180	11.83	Great Central Railway	OP
08123	D3190	03.84	Cholsey & Wallingford Railway	OP
08133	D3201	09.80	Co-Steel Sheerness	SU
08157	D3225	04.77	RFS(E) Ltd. (Hire Locomotive 025)	OP
08164	D3232	03.86	RFS(E) Ltd. (Hire Locomotive 002)	OP
08168	D3236	03.88	ABB Transportation, York	OP

08177	D3245	10.88	ABB Transportation, Crewe	
	D3255	12.72	Brighton Railway Museum	SU
	D3261	12.72	Brighton Railway Museum	SU
08195	D3265	09.83	Llangollen Railway	OP
08202	D3272	04.89	Potter Group, Ely	OP
08216	D3286	11.80	Co-Steel Sheerness	SU
08220	D3290	03.86	Steamtown Railway Centre	OP
08238	D3308	03.84	Dean Forest Railway	OP
08266	D3336	03.85	Keighley & Worth Valley Railway	OP
08288	D3358	01.83	Mid-Hants Railway	OP
08292	D3362	05.84	Deanside Transit	SP
08296	D3366	10.88	ABB Transportation, Crewe (ABB 001)	OP
08308	D3378	02.92	South Yorkshire Railway	SU
08320	D3390	12.82	English China Clays, Burngullow	OP
08331	D3401	03.88	RFS(E) Ltd. (Hire Locomotive 001)	OP
08345	D3415	10.83	Deanside Transit	OP
08350	D3420	01.84	Cheddleton Railway Centre	OP
08359	D3429	01.84	Peak Rail	OP
08375	D3460	11.91	The Railway Age, Crewe	OP
08377	D3462	06.83	Dean Forest Railway	OP
08398	D3513	07.85	English China Clays, Rocks Driers, Bugle	OP
08423	D3538	11.88	Trafford Park Estates (hired to Locotec)	OP
08436	D3551	01.92	South Yorkshire Railway	SU
08443	D3558	07.87	Bo'ness & Kinneil Railway	UR
08444	D3559	11.86	Bodmin & Wenford Railway	OP
08470	D3585	03.86	ABB Transportation, Crewe	OP
08471	D3586	09.85	Severn Valley Railway	OP
08476	D3591	09.85	Swanage Railway	OP
08479	D3594	11.91	East Lancashire Railway	OP
08490	D3605	12.85	Strathspey Railway	OP
08502	D3657	09.88	ICI, Wilton	OP
08503	D3658	09.88	ICI, Wilton	OP
08556	D3723	07.90	North Yorkshire Moors Railway	OP
08590	D3757	09.93	Midland Railway Centre	OP
08596	D3763	03.77	RFS(E) Ltd. (Hire Locomotive 006)	OP
08598	D3765	11.86	PD Fuels, Gwaun-cae-Gurwen	OP
08602	D3769	03.86	ABB Transportation, Derby C & W	OP
08604	D3771	07.93	Didcot Railway Centre	
08613	D3780	12.93	Trafford Park Estates	OP
08615	D3782	12.93	Trafford Park Estates	OP
08631	D3798	12.92	Mid Norfolk Railway	UR
08650	D3817	08.89	Foster Yeoman, Isle of Grain	OP
08652	D3819	06.92	Foster Yeoman, Merehead	OP
08669	D3836	05.89	Trafford Park Estates	OP

08678	D3845	09.88	Steamtown Railway Centre	OP
08704	D3871	11.90	Nene Valley Railway	OP
08728	D3896	09.87	Deanside Transit	OP
08736	D3904	09.87	Deanside Transit	OP
08743	D3911	03.93	RFS(E) Ltd. (Hire Locomotive 024)	OP
08764	D3932	05.88	RFS(E) Ltd. (Hire Locomotive 003)	OP
08767	D3935	01.94	North Norfolk Railway	OP
08769	D3937	04.89	Fire Services Training Centre	SU
08772	D3940	01.94	East Anglian Railway Museum	UR
08774	D3942	09.88	A.V. Dawson, Middlesborough	OP
08785	D3953	03.89	RFS(E) Ltd. (Hire Locomotive 004)	OP
08788	D3956	01.94	Great Central Railway	OP
08802	D3970	09.93	Cholsey & Wallingford Railway	
08809	D3984	12.93	Otis Euro-Transrail, Salford	OP
08816	D3984	02.86	South Yorkshire Railway	OP
08846	D4014	10.89	ABB Transportation, Crewe	OP
08850	D4018	12.92	West Somerset Railway	OP
08868	D4036	12.92	East Lancashire Railway	OP
08870	D4038	05.93	South Yorkshire Railway	OP
08871	D4039	10.90	Humberside Sea & Land Services, Grimsby	OP
08874	D4042	02.92	RFS(E) Ltd. (Hire Locomotive 023)	OP
08885	D4115	05.93	Great Central Railway, Ruddington	SU
08915	D4145	12.93	Gloucestershire Warwickshire Railway	OP
08936	D4166	12.92	South Yorkshire Railway	SU
08937	D4167	12.93	Camas Aggregates, Meldon Quarry	OP
08943	D4167	07.88	ABB Transportation, Crewe (ABB 002)	OP

CLASS 09   0-6-0 DE

Built: 1959 by BR at Darlington Works.
Engine: English Electric 6KT of 297 kW at 680 rpm.
Length over buffers: 8.915 m.   **Weight:** 47.10 t.
Extreme width: 2.915 m.   **Wheel Diameter:** 1372 mm.
Extreme height: 3.877 m.   **Maximum Permitted Speed:** 27.5 mph.

| 09002 | D3666 | 09.92 | South Devon Railway | UR |

CLASS 10   0-6-0 DE

Built: 1957-62 by BR at Darlington Works.
Engine: Lister Blackstone ER6T of 261 kW at 750 rpm.
Length over buffers: 8.915 m.   **Weight:** 49.00 t.
Extreme width: 2.915 m.   **Wheel Diameter:** 1372 mm.
Extreme height: 3.877 m.   **Maximum Permitted Speed:** 20 mph.

D3452	07.68	Bodmin & Wenford Railway	OP
D3476	06.68	South Yorkshire Railway	SU
D3489	04.68	Felixstowe Dock & Railway Company	OP
D3639	07.69	Boke Railway, Guinea	EX
D3649	07.69	Boke Railway, Guinea	EX
D4067	12.70	Great Central Railway	OP
D4092	09.68	South Yorkshire Railway	SU

CLASS 11   0-6-0 DE

Built: 1949-52 by BR at Derby Locomotive or Darlington Works.
Engine: English Electric 6KT of 260 kW at 680 rpm.
Length over buffers: 8.877 m.   **Weight:** 47.25 t.
Extreme width: **Wheel Diameter:** 1232 mm.
Extreme height: **Maximum Permitted Speed:** 20 mph.
Note: * Currently on loan to Cobra, Wakefield.

12049	10.71	Day Aggregates, Brentford	OP
12052	06.71	Scottish Industrial Railway Museum	IE
12061	10.71	Gwili Railway/Rheilffordd Gwili	OP
12071	10.71	South Yorkshire Railway	SU
12074	01.72	South Yorkshire Railway	SU
12077	10.71	Midland Railway Centre	OP
12082	10.71	South Yorkshire Railway*	OP
12083	10.71	TILCON, Grassington	SS
12088	05.71	South Yorkshire Railway	SU
12093	05.71	Scottish Industrial Railway Centre	UR
12098	02.71	North Tyneside Steam Railway	OP
12099	07.71	Severn Valley Railway	UR
12131	07.71	North Norfolk Railway	OP

CLASS 12   0-6-0 DE

Built: 1949 by BR at Ashford Works.
Engine: English Electric 6KT of 260 kW at 680 rpm.
Length over buffers: 8.985 m.   **Weight:** 48.15 t.
Extreme width: **Wheel Diameter:** 1372 mm.
Extreme height: **Maximum Permitted Speed:** 27.5 mph.

| 15224 | 10.71 | Lavender Line | OP |

CLASS 14   0-6-0 DH

Built: 1964-65 by BR at Swindon Works.
Engine: Paxman Ventura 6YJXL of 485 kW at 1500 rpm.

Length over buffers: 10.541 m.		**Weight:** 50.00 t.	
Extreme width:		**Wheel Diameter:** 1219 mm.	
Extreme height:		**Maximum Permitted Speed:** 40 mph.	
D9500	04.69	South Yorkshire Railway	SU
D9502	04.69	South Yorkshire Railway	SU
D9504	04.68	Kent & East Sussex Steam Railway	SU
D9505	04.68	Maldegem, Belgium	EX
D9513	03.68	Embsay Steam Railway	SS
D9515	04.68	Charmartin, Madrid	SU
D9516	04.68	Nene Valley Railway	OP
D9518	04.69	Rutland Railway Museum	SU
D9520	04.68	Rutland Railway Museum	OP
D9521	04.69	Swanage Railway	SU
D9523	04.68	Nene Valley Railway	OP
D9524	04.69	Bo'ness & Kinneil Railway	UR
D9525	04.68	Kent & East Sussex Steam Railway	OP
D9526	11.68	West Somerset Railway	UR
D9529	04.68	Nene Valley Railway	OP
D9531	12.67	East Lancashire Railway	OP
D9534	04.68	Milan, Italy	EX
D9537	04.68	Gloucestershire Warwickshire Railway	SU
D9539	04.68	Gloucestershire Warwickshire Railway	OP
D9548	04.68	Charmartin, Madrid	SU
D9549	04.68	Charmartin, Madrid	SU
D9551	04.68	West Somerset Railway	OP
D9553	04.68	Gloucestershire Warwickshire Railway	OP
D9555	04.68	Rutland Railway Museum	UR

CLASS 15   Bo-Bo DE

Built: 1960 by Clayton Equipment Company, Hatton.
Engine: Paxman 16YHXL of 597 kW.

Length over buffers: 12.887 m.		**Weight:** 68.00 t.	
Extreme width: 2.794 m.		**Wheel Diameter:** 991 mm.	
Extreme height: 3.810 m.		**Maximum Permitted Speed:** 60 mph.	

Subsequent Number: ADB 968001 in Departmental Service.

D8233	02.69	The Railway Age, Crewe	UR

CLASS 17   Bo-Bo DE

Built: 1964 by Clayton Equipment Company, Hatton.
Engines: Two Paxman 6ZHXL of 336 kW each at 1500 rpm.
Length over buffers: 15.431 m.   **Weight:** 68.00 t.

Extreme width: 2.680 m.　　　**Wheel Diameter:** 1003 mm.
Extreme height: 3.861 m.　　　**Maximum Permitted Speed:** 60 mph.
　　　D8568　　10.71　　Chinnor & Princes Risborough Railway　　OP

CLASS 20 　　　　　　　　　　　　　　　　　　Bo-Bo DE

Built: 1957-68 by English Electric, Vulcan Foundry, Newton-le-Willows or Robert Stephenson & Hawthorn, Darlington.
Engine: English Electric 8SVT Mk. II of 746 kW at 850 rpm.
Length over buffers: 14.259 m.　　　**Weight:** 72.05 (* 72.70, † 71.85) t.
Extreme width: 2.667 m.　　　　　　**Wheel Diameter:** 1092 mm.
Extreme height: 3.851 m.　　　　　 **Maximum Permitted Speed:** 75 mph.
Notes: 20172 has also carried 20305. 20194 has also carried 20307.

20001*	D8001	04.88	Midland Railway Centre	UR
20020*	D8020	10.90	Bo'ness & Kinneil Railway	UR
20031*	D8031	09.90	Keighley & Worth Valley Railway	OP
20035*	D8035	07.91	CFD, Autun, France. (2001)	OP
20041*	D8041	11.88	Hunslet-Barclay Ltd. (20901)	ML
20042*	D8042	06.91	The Railway Age, Crewe	SU
20047*	D8047	09.91	RFS Locomotives. (2004)	SS
20048*	D8048	11.90	Peak Rail	OP
20050*	D8000	12.80	National Railway Museum	UR
20056†	D8056	10.90	Caledonian Railway	OP
20060†	D8060	11.88	Hunslet-Barclay Ltd. (20902)	ML
20063†	D8063	07.91	CFD, Autun, France. (2002)	OP
20069†	D8069	05.91	Mid-Norfolk Railway	OP
20083†	D8083	01.89	Hunslet-Barclay Ltd. (20903)	ML
20084†	D8084	09.91	RFS Locomotives. (2002)	SS
20088†	D8088	09.91	RFS Locomotives. (2017)	SS
20094†	D8094	06.93	Class 20 Workgroup, GCR Ruddington	UR
20095†	D8095	02.91	RFS Locomotives. (2020)	SS
20096†	D8096	06.91	South Yorkshire Railway	UR
20098†	D8098	06.91	Great Central Railway	OP
20101†	D8101	11.88	Hunslet-Barclay Ltd. (20904)	ML
20102†	D8102	09.91	RFS Locomotives. (2008)	SS
20105†	D8105	09.91	RFS Locomotives. (2016)	SS
20107†	D8107	01.91	Locotec Hire Locomotive	OP
20108†	D8108	09.91	RFS Locomotives. (2001)	SS
20110†	D8110	09.91	South Devon Railway	OP
20113†	D8113	09.91	RFS Locomotives. (2003)	SS
20120†	D8120	09.91	RFS Locomotives. (2009)	SS
20127†	D8127	01.91	RFS Locomotives. (2018)	SS
20133	D8133	09.91	RFS Locomotives. (2005)	SS

Class 20 20098 at the Great Central Railway's Loughborough depot on 09.08.92.

(John Stretton)

20135	D8135	07.93	Class 20 Workgroup, GCR Ruddington	SU
20137	D8137	12.93	Gloucestershire Warwickshire Railway	SU
20139	D8139	07.91	CFD, Autun, France. (2003)	OP
20142	D8142	07.93	Llangollen Railway	OP
20145	D8145	05.91	RFS Locomotives. (2019)	SS
20159	D8159	09.91	RFS Locomotives. (2010)	SS
20166	D8166	05.91	Bodmin & Wenford Railway	OP
20175	D8175	09.91	RFS Locomotives. (2007)	SS
20188	D8188	01.90	The Railway Age, Crewe	OP
20189	D8189	09.90	M.C. Metal Processing, Glasgow	OP
20194	D8194	09.91	RFS Locomotives. (2006)	SS
20197	D8197	11.91	Bodmin & Wenford Railway	SU
20206	D8306	04.91	Gloucestershire Warwickshire Railway	OP
20209	D8309	11.88	Hunslet-Barclay Ltd, Kilmarnock	SU
20214	D8314	09.93	Lakeside & Haverthwaite Railway	SU
20219	D8319	01.89	Hunslet-Barclay Ltd. (20906)	ML
20225	D8325	01.89	Hunslet-Barclay Ltd. (20905)	ML
20227	D8327	10.90	Midland Railway Centre	OP
20228†	D8128	07.91	CFD, Autun, France. (2004)	OP

CLASS 24   Bo-Bo DE

Built: 1959-60 by BR at Crewe Works.
Engine: Sulzer 6LDA28A of 866 kW at 750 rpm.
Length over buffers: 15.392 m.   **Weight:** 73.00 (* 78.70) t.
Extreme width: 2.769 m.   **Wheel Diameter:** 1143 mm.
Extreme height: 3.861 m.   **Maximum Permitted Speed:** 75 mph.
Subsequent Numbers: 24054 later became TDB 968008 in Departmental Service. 24061 later became TDB 968008 and then 97201 in Departmental Service.

24032*	D5032	07.76	North Yorkshire Moors Railway	OP
24054	D5054	07.76	East Lancashire Railway	OP
24061	D5061	08.75	North Yorkshire Moors Railway	OP
24081	D5081	10.80	Llangollen Railway	OP

CLASS 25   Bo-Bo DE

Built: 1963-67 by BR at Darlington or Derby Locomotive Works, or Beyer Peacock, Manchester.
Engine: Sulzer 6LDA28-B of 930 kW at 750 rpm.
Length over buffers: 15.392 m.   **Weight:** 71.45 (* 73.75) t.
Extreme width: 2.734 m.   **Wheel Diameter:** 1143 mm.
Extreme height: 3.861 m.   **Maximum Permitted Speed:** 90 mph.

(John Stretton)

Class 24 24054 at Crewe Basford Hall Yard on 21.08.94.

Subsequent Numbers: 25901 (25262), 25904 (25283), 25909 (25309), 25912 (25322).

25035*	D5185	03.87	Northampton & Lamport Railway	OP
25057*	D5207	03.87	North Norfolk Railway	OP
25059*	D5209	03.87	Keighley & Worth Valley Railway	UR
25067*	D5217	12.82	Mid-Hants Railway	UR
25072*	D5222	12.85	The Railway Age, Crewe	UR
25173	D7523	03.87	The Railway Age, Crewe	OP
25185	D7535	11.84	Paignton & Dartmouth Steam Railway	OP
25191	D7541	03.87	North Yorkshire Moors Railway	OP
25205	D7555	09.86	P.Waterman	SB
25235*	D7585	03.85	Bo'ness & Kinneil Railway	OP
25244*	D7594	07.86	Nene Valley Railway	UR
25262	D7612	03.87	East Lancashire Railway	UR
25265	D7615	03.87	Peak Rail	OP
25278	D7628	03.87	North Yorkshire Moors Railway	OP
25279	D7629	03.87	Llangollen Railway	OP
25283	D7633	03.87	Severn Valley Railway	OP
25309	D7659	09.86	The Railway Age, Crewe	OP
25313	D7663	03.87	Llangollen Railway	OP
25321	D7671	09.86	Midland Railway Centre	OP
25322	D7672	09.91	Cheddleton Railway Centre	OP

CLASS 26   Bo-Bo DE

Built: 1958-59 by Birmingham Railway Carriage & Wagon Company, Smethwick.
Engine: Sulzer 6LDA28-B of 870 kW at 750 rpm.
Length over buffers: 15.469 m.   **Weight:** 74.50 (* 75.00) t.
Extreme width: 2.692 m.   **Wheel Diameter:** 1092 mm.
Extreme height: 3.861 m.   **Maximum Permitted Speed:** 75 mph.

26002	D5302	10.92	Strathspey Railway	SB
26004	D5304	11.92	Bo'ness & Kinneil Railway	SB
26010	D5310	12.92	Northampton & Lamport Railway	SB
26011	D5311	11.92	Peak Rail	SB
26014	D5314	10.92	Caledonian Railway	SB
26024	D5324	10.92	Bo'ness & Kinneil Railway	SB
26026	D5326	11.92	P. Waterman	SB
26035	D5335	12.92	Bodmin & Wenford Railway	SB
26038	D5338	10.92	South Yorkshire Railway	SB
26041	D5341	11.92	Mid-Norfolk Railway	SB
26043	D5343	01.93	Gloucestershire Warwickshire Railway	SB

CLASS 27   Bo-Bo DE

Built: 1961-62 by Birmingham Railway Carriage & Wagon Company, Smethwick.
Engine: Sulzer 6LDA28-B of 930 kW at 750 rpm.
Length over buffers: 15.469 m.   **Weight:** 73.30 (* 71.20) t.
Extreme width: 2.692 m.   **Wheel Diameter:** 1092 mm.
Extreme height: 3.861 m.   **Maximum Permitted Speed:** 90 mph.
Subsequent Number: 27024 later became ADB 968028 in Departmental Service.
Notes: 27106 has also carried 27050. 27112 has also carried 27056. 27205 has
also carried 27123 and 27059. 27212 has also carried 27103 & 27066.

27001	D5347	07.87	Bo'ness & Kinneil Railway	OP
27005	D5351	07.87	Bo'ness & Kinneil Railway	OP
27007	D5353	01.85	Mid-Hants Railway	OP
27024*	D5370	07.87	Caledonian Railway	OP
27106	D5394	07.87	Strathspey Railway	OP
27112	D5401	02.87	Northampton & Lamport Railway	OP
27205	D5410	07.87	Birmingham Railway Museum	OP
27212	D5386	07.87	North Norfolk Railway	OP

CLASS 28   Co-Bo DE

Built: 1958 by Metropolitan Vickers, Stockton on Tees.
Engine: Crossley HSTV8 of 896 kW at 625 rpm.
Length over buffers: 17.259 m.   **Weight:** 97.00 t.
Extreme width: 2.807 m.   **Wheel Diameter:** 1003 mm.
Extreme height: 3.870 m.   **Maximum Permitted Speed:** 75 mph.
Subsequent Numbers: SI5705, TDB 968006 in Departmental Service.

	D5705	09.68	Peak Rail	SU

CLASS 31   A1A-A1A DE

Built: 1957 by Brush Traction, Loughborough.
Engine: English Electric 12SVT of 1100 kW (1470hp) at 850 rpm.
Length over buffers: 17.297 m.   **Weight:** 104.00 t.
Extreme width: 2.667 m.   **Powered Wheel Diameter:** 1092 mm.
Extreme height: 3.848 m.   **Unpowered Wheel Diameter:** 1003 mm.
Maximum Permitted Speed: 75 mph.

31018	D5500	07.76	Steamtown Railway Centre	SU
31101	D5518	01.93	East Lancashire Railway	OP
31123	D5541	03.93	Gloucestershire Warwickshire Railway	OP
31162	D5580	05.92	Midland Railway Centre	OP
31286	D5818	12.91	Bodmin & Wenford Railway	SB

CLASS 33   Bo-Bo DE

Built: 1960-62 by Birmingham Railway Carriage & Wagon Company, Smethwick.
Engine: Sulzer 8LDA28 of 1160 kW at 750 rpm.
Length over buffers: 15.240 m.   **Weight:** 76.45 (* 76.25) t.
Extreme width: 2.743 (* 2.642) m.   **Wheel Diameter:** 1092 mm.
Extreme height: 3.861 m.   **Maximum Permitted Speed:** 85 mph.
Subsequent Number: 33115 has also carried 83301 as a loco-hauled test vehicle.

33034	D6552	02.88	Ministry of Defence, Ludgershall	UR
33056	D6574	02.91	Cheddleton Railway Centre	SU
33102	D6513	11.92	Cheddleton Railway Centre	SU
33110	D6527	09.92	Bodmin & Wenford Railway	SU
33111	D6528	06.91	St. Leonards Railway Engineering	UR
33114	D6532	02.93	Ministry of Defence, Ludgershall	SP
33115	D6533	05.89	St. Leonards Railway Engineering	UR
33203	D6588	04.91	South Yorkshire Railway	SU

CLASS 35   B-B DH

Built: 1962-63 by Beyer Peacock, Manchester.
Engine: Maybach MD870 of 1300 kW at 1500 rpm.
Length over buffers: 15.761 m.   **Weight:** 74.00 t.
Extreme width: 2.654 m.   **Wheel Diameter:** 1143 mm.
Extreme height: 3.926 m.   **Maximum Permitted Speed:** 90 mph.

D7017	03.75	West Somerset Railway	OP
D7018	03.75	West Somerset Railway	OP
D7029	02.75	North Yorkshire Moors Railway	SU
D7076	05.73	East Lancashire Railway	OP

CLASS 37   Co-Co DE

Built: 1960-65 by English Electric, Vulcan Foundry, Newton-le-Willows or Robert Stephenson & Hawthorn, Darlington.
Engine: English Electric 12CSVT of 1300 kW at 850 rpm.
Length over buffers: 18.745 m.   **Weight:** 102.80-108.40 t.
Extreme width: 2.743 m.   **Wheel Diameter:** 1092 mm.
Extreme height: 3.937 m.   **Maximum Permitted Speed:** 90 mph.

37029	D6729	12.93	The Railway Age, Crewe	UR
37032	D6732	03.94	North Norfolk Railway	UR
37190	D6890	07.93	Midland Railway Centre	UR
37215	D6915	07.93	Gloucestershire Warwickshire Railway	SU

Class 35 D7018 double heads with Class 52 D1035 at Didcot Railway Centre on 05.05.90.   (John Stretton)

CLASS 40   1Co-Co1 DE

Built: 1958-61 by English Electric, Vulcan Foundry, Newton-le-Willows or Robert Stephenson & Hawthorn, Darlington.
Engine: English Electric 16SVT Mk. II of 1493 kW at 850 rpm.
Length over buffers: 21.184 m.   **Weight:** 133.00 t.
Extreme width: 2.743 m.   **Powered Wheel Diameter:** 1143 mm.
Extreme height: 3.921 m.   **Unpowered Wheel Diameter:** 914 mm.
Maximum Permitted Speed: 90 mph.
Subsequent Numbers: 97407 (40012), 97408 (40118), 97406 (40135) in Departmental Service.

40012	D212	02.85	Midland Railway Centre	OP
40013	D213	01.85	South Yorkshire Railway	UR
40106	D306	04.83	Nene Valley Railway	OP
40118	D318	02.85	Birmingham Railway Museum	UR
40122	D200	04.88	National Railway Museum	OP
40135	D335	01.85	East Lancashire Railway	OP
40145	D345	06.83	East Lancashire Railway	OP

CLASS 42   B-B DH

Built: 1960-61 by BR at Swindon Works.
Engines: Two Maybach MD650 of 820 kW each at 1530 rpm.
Length over buffers: 18.288 m.   **Weight:** 78.00 t.
Extreme width: 2.654 m.   **Wheel Diameter:** 1003 mm.
Extreme height: 3.899 m.   **Maximum Permitted Speed:** 90 mph.

D821	12.72	Severn Valley Railway	OP
D832	12.72	East Lancashire Railway	UR

CLASS 43   Bo-Bo DE

Built: 1972 by British Rail Engineering Ltd., Crewe Works.
Engine: Paxman Valenta 12RP200L of 1680 kW at 1500 rpm.
Length over buffers: 17.145 m.   **Weight:** 66.00 t.
Extreme width: 2.743 m.   **Wheel Diameter:** 1016 mm.
Extreme height: 3.912 m.   **Maximum Permitted Speed:** 125 mph.
Original Number: 41001.
Subsequent Number: ADB 975812 in Departmental Service.

43000	11.76	National Railway Museum	IE

CLASS 44   1Co-Co1 DE

Built: 1959 by BR at Derby Locomotive Works.
Engine: Sulzer 12LDA28A of 1716 kW at 750 rpm.
Length over buffers: 20.701 m.   **Weight:** 133.00 t.
Extreme width: 2.783 m.   **Powered Wheel Diameter:** 1143 mm.
Extreme height: 3.915 m.   **Unpowered Wheel Diameter:** 914 mm.
Maximum Permitted Speed: 90 mph.

44004	D4	11.80	Great Central Railway	OP
44008	D8	11.80	Peak Rail	OP

CLASS 45   1Co-Co1 DE

Built: 1960-62 by BR at Derby Locomotive or Crewe Works.
Engine: Sulzer 12LDA28B of 1866 kW at 750 rpm.
Length over buffers: 20.701 m.   **Weight:** 136.10 t.
Extreme width: 2.783 m.   **Powered Wheel Diameter:** 1143 mm.
Extreme height: 3.915 m.   **Unpowered Wheel Diameter:** 914 mm.
Maximum Permitted Speed: 90 mph.

45041	D53	06.86	The Railway Age, Crewe	SU
45060	D100	12.85	Peak Rail	OP
45105	D86	05.87	Peak Rail	UR
45108	D120	08.87	The Railway Age, Crewe	OP
45112	D61	05.87	East Lancashire Railway	UR
45118	D67	05.87	Northampton & Lamport Railway	UR
45125	D123	05.87	BR Dairycoates Depot, Hull	UR
45132	D22	05.87	Mid-Hants Railway	OP
45133	D40	05.87	Midland Railway Centre	OP
45135	D99	03.87	Peak Rail	OP
45149	D135	09.87	The Railway Age, Crewe	UR

CLASS 46   1Co-Co1 DE

Built: 1961-63 by BR at Derby Locomotive Works.
Engine: Sulzer 12LDA28B of 1866 kW at 750 rpm.
Length over buffers: 20.701 m.   **Weight:** 138.00 t.
Extreme width: 2.783 m.   **Powered Wheel Diameter:** 1143 mm.
Extreme height: 3.915 m.   **Unpowered Wheel Diameter:** 914 mm.
Maximum Permitted Speed: 90 mph.
Subsequent Numbers: 97403 (46035), 97404 (46045) in Departmental Service.

46010	D147	11.84	Llangollen Railway	UR
46035	D172	11.84	The Railway Age, Crewe	ML
46045	D182	11.84	Midland Railway Centre	OP

CLASS 47 Co-Co DE

Built: 1962-65 by Brush Traction, Loughborough or BR, Crewe Works.
Engine: Sulzer 12LDA28C of 2052 kW at 800 rpm.

Length over buffers: 19.355 m.		**Weight:** 116.90 (* 115.35) t.		
Extreme width: 2.794 m.		**Wheel Diameter:** 1143 mm.		
Extreme height: 3.896 m.		**Maximum Permitted Speed:** 95 mph.		

47105	D1693	12.93	Gloucestershire Warwickshire Railway	OP
47117	D1705	07.91	East Lancashire Railway	OP
47192*	D1842	05.88	The Railway Age, Crewe	OP
47401	D1500	06.92	Midland Railway Centre	OP
47402	D1501	06.92	East Lancashire Railway	SS
47417	D1516	02.92	Midland Railway Centre	SB
47449*	D1566	05.93	Birmingham Railway Museum	OP
47488	D1713	10.93	The Railway Age, Crewe	UR

CLASS 50 Co-Co DE

Built: 1967-68 by English Electric, Vulcan Foundry, Newton-le-Willows.
Engine: English Electric 16SVT of 2015 kW at 850 rpm.

Length over buffers: 20.879 m.		**Weight:** 115.05 t.		
Extreme width: 2.775 m.		**Wheel Diameter:** 1092 mm.		
Extreme height: 3.956 m.		**Maximum Permitted Speed:** 100 mph.		

Subsequent Numbers: 50149 (50049).

50002	D402	09.91	Paignton & Dartmouth Steam Railway	OP
50007	D407	03.94	Midland Railway Centre	OP
50008	D408	06.92	The Railway Age, Crewe	OP
50015	D415	06.92	East Lancashire Railway	SU
50017	D417	09.91	The Railway Age, Crewe	UR
50019	D419	09.90	Spa Valley Railway	UR
50021	D421	04.90	MOD Bicester Military Railway	OP
50026	D426	12.90	MOD Bicester Military Railway	SU
50027	D427	07.91	North Yorkshire Moors Railway	OP
50031	D431	08.91	St. Leonards Railway Engineering	OP
50033	D433	06.92	National Railway Museum	OP
50035	D435	08.90	St. Leonards Railway Engineering	UR
50042	442	10.90	Bodmin & Wenford Railway	OP
50043	443	02.91	Birmingham Railway Museum	IE
50044	444	01.91	St. Leonards Railway Engineering	UR
50049	449	08.91	West Somerset Railway	OP
50050	D400	03.94	St. Leonards Railway Engineering	OP

CLASS 52　　　　　　　　　　　　　　　　　　　　　C-C DH

Built: 1962-63 by BR at Swindon or Crewe Works.
Engines: Two Maybach MD655 of 1007 kW each at 1500 rpm.
Length over buffers: 20.726 m.　　　**Weight:** 108.00 t.
Extreme width: 2.743 m.　　　　　　**Wheel Diameter:** 1092 mm.
Extreme height: 3.959 m.　　　　　　**Maximum Permitted Speed:** 90 mph.
Note: D1010 is preserved as D1035 "Western Yeoman"

D1010	02.77	West Somerset Railway	OP
D1013	02.77	Severn Valley Railway	UR
D1015	12.76	BR Old Oak Common Depot	UR
D1023	02.77	National Railway Museum	UR
D1041	02.77	East Lancashire Railway	OP
D1048	02.77	The Railway Age, Crewe	UR
D1062	08.74	Severn Valley Railway	OP

CLASS 55　　　　　　　　　　　　　　　　　　　　Co-Co DE

Built: 1961 by English Electric, Vulcan Foundry, Newton-le-Willows.
Engines: Two Napier Deltic D18-25 of 1230 kW each at 1500 rpm.
Length over buffers: 21.184 m.　　　**Weight:** 99.90 t.
Extreme width: 2.680 m.　　　　　　**Wheel Diameter:** 1092 mm.
Extreme height: 3.937 m.　　　　　　**Maximum Permitted Speed:** 100 mph.

55002	D9002	01.82	North Tyneside Steam Railway	CE
55009	D9009	01.82	ICI, Wilton	UR
55015	D9015	01.82	Midland Railway Centre	OP
55016	D9016	12.81	BR Old Oak Common Depot	OP
55019	D9019	12.81	Great Central Railway	OP
55022	D9000	01.82	BR Old Oak Common Depot	UR

CLASS 71　　　　　　　　　　　　　　　　　　　　　Bo-Bo

Built: 1959 by BR at Doncaster Works.
Supply System: 660-750 V dc third rail or overhead.
Traction Motors: Electric. Four English Electric type 532.
Continuous Rating: 1716 kW at 69.6 mph.
Length over buffers: 15.418 m.　　　**Weight:** 77.00 t.
Extreme width: 2.821 m.　　　　　　**Wheel Diameter:** 1219 mm.
Extreme height: 3.988 m.　　　　　　**Maximum Permitted Speed:** 90 mph.

71001	E5001	11.77	National Railway Museum	ML

CLASS 76   Bo+Bo

Built: 1951 by BR at Gorton Works.
Supply System: 1500 v dc overhead.
Traction Motors: Four Metropolitan Vickers type 486.
Continuous Rating: 970 kW at 56 mph.
Length over buffers: 15.392 m.   **Weight:** 87.30 t.
Extreme width: 2.743 m.   **Wheel Diameter:** 1270 mm.
Extreme height: 4.191 m.   **Maximum Permitted Speed:** 65 mph.
Original Number: 26020.

76020	E26020	07.81	National Railway Museum	IE

CLASS 77   Co-Co

Built: 1953-54 by BR at Gorton Works.
Supply System: 1500 V dc overhead.
Traction Motors: Six Metropolitan Vickers type 146.
Continuous Rating: 78 kN at 23 mph.
Length over buffers: 17.983 m.   **Weight:** 102.50 t.
Extreme width: **Wheel Diameter:** 1092 mm.
Extreme height: **Maximum Permitted Speed:** 90 mph.
Original Numbers: 27000/01/03 respectively.
Subsequent Numbers: NS 1502/05/01 respectively.

E27000	09.68	Midland Railway Centre	CE
E27001	09.68	Museum of Science & Industry, Manchester	CE
E27003	09.68	Rotterdam, The Netherlands.	OP

CLASS 81   Bo-Bo

Built: 1960 by Birmingham Railway Carriage & Wagon Company, Smethwick.
Supply System: 25 kV ac 50 Hz overhead.
Traction Motors: Four AEI type 189.
Continuous Rating: 2388 kW at 71 mph.
Length over buffers: 17.221 m.   **Weight:** 80.00 t.
Extreme width: 2.648 m.   **Wheel Diameter:** 1219 mm.
Extreme height: 3.977 m.   **Maximum Permitted Speed:** 100 mph.

81002	E3003	10.90	The Railway Age, Crewe	CE

CLASS 82   Bo-Bo

Built: 1961 by Beyer Peacock, Manchester.
Supply System: 25 kV ac 50 Hz overhead.

A rare view of Class 76 E26020 at Swindon on 19.05.90.

(John Stretton)

Traction Motors: Four AEI type 189.
Continuous Rating: 2463 kW at 73 mph.
Length over buffers: 17.069 m.   **Weight:** 77.45 t.
Extreme width: 2.667 m.   **Wheel Diameter:** 1219 mm.
Extreme height: 3.977 m.   **Maximum Permitted Speed:** 100 mph.

82008	E3054	12.87	The Railway Age, Crewe	IE

CLASS 83   Bo-Bo

Built: 1961 by English Electric, Vulcan Foundry, Newton-le-Willows.
Supply System: 25 kV ac 50 Hz overhead.
Traction Motors: Four English Electric type 535A.
Length over buffers: 16.002 m.   **Weight:** 74.50 t.
Extreme width: 2.661 m.   **Wheel Diameter:** 1219 mm.
Extreme height: 3.978 m.   **Maximum Permitted Speed:** 100 mph.

83012	E3035	03.89	The Railway Age, Crewe	IE

CLASS 84   Bo-Bo

Built: 1960 by North British Locomotive Company, Glasgow.
Supply System: 25 kV ac 50 Hz overhead.
Traction Motors: Four GEC type WT501.
Continuous Rating: 2313 kW at 66 mph.
Length over buffers: 16.320 m.   **Weight:** 77.00 t.
Extreme width: 2.648 m.   **Wheel Diameter:** 1219 mm.
Extreme height: 3.978 m.   **Maximum Permitted Speed:** 100 mph.
Note: On extended loan from British Rail.

84001	E3036	01.79	National Railway Museum	IE

CLASS 85   Bo-Bo

Built: 1961 by BR at Doncaster Works.
Supply System: 25 kV ac 50 Hz overhead.
Traction Motors: Four AEI type 189.
Continuous Rating: 2388 kW at 71 mph.
Length over buffers: 17.221 m.   **Weight:** 79.65 t.
Extreme width: 2.648 m.   **Wheel Diameter:** 1219 mm.
Extreme height: 3.978 m.   **Maximum Permitted Speed:** 100 mph.
Subsequent Number: 85101.

85006	E3061	11.92	The Railway Age, Crewe	SU

CLASS 89   Co-Co

Built: 1986 by BREL, Crewe.
Supply System: 25 kV ac 50 Hz overhead.
Traction Motors: Six Brush type TM2201A.
Continuous Rating: 3200 kW at 91.8 mph.
Length over buffers: 19.798 m.   **Weight:** 102.25 t.
Extreme width: 2.736 m.   **Wheel Diameter:** 1070 mm.
Extreme height: 3.977 m.   **Maximum Permitted Speed:** 125 mph.

89001	07.92	Midland Railway Centre	CE

CLASS 97/6   0-6-0 DE

Built: 1952 by Ruston & Hornsby, Lincoln.
Engine: Ruston 6VPH of 123 kW.
Length over buffers: 7.620 m.   **Weight:** 30.20 t.
Extreme width: 2.565 m.   **Wheel Diameter:** 978 mm.
Extreme height: 3.353 m.   **Maximum Permitted Speed:** 20 mph.

97650	PWM650	04.87	BSC Scunthorpe	SU

CLASS 98/1   0-6-0 DH

Built: 1987 by Brecon Mountain Railway, Pant.
Gauge: 597 mm.
Engine: Caterpillar 3304T of 105 kW.
Length over buffers: 5.029 m.   **Weight:** 12.75 t.
Extreme width: **Wheel Diameter:** 610 mm.
Extreme height: **Maximum Permitted Speed:** 15 mph.

10	04.89	Vale of Rheidol Railway	OP

1.2 LOCOMOTIVES NOT CLASSIFIED UNDER THE TOPS SYSTEM

UNCLASSIFIED   A1A-A1A GTE

Built: 1949 by Brown Boveri, Switzerland.
Turbine: Brown Boveri of 1866 kW.
Length over buffers: 19.215 m.   **Weight:** 119.15 t.
Extreme width: 2.807 m.   **Powered Wheel Diameter:** 1232 mm.
Extreme height: 4.064 m.   **Unpowered Wheel Diameter:** 965 mm.

Maximum Permitted Speed: 90 mph.
18000   02.60   The Railway Age, Crewe   IE

UNCLASSIFIED   0-6-0 DM

Built: 1961 by Hudswell Clarke, Leeds.
Engine: Gardner 8L3 of 152 kW at 1200 rpm.
Length over buffers: 8.007 m.   **Weight:** 34.20 t.
Extreme width: **Wheel Diameter:** 1067 mm.
Extreme height: **Maximum Permitted Speed:** 24 mph.
D2511   12.67   Keighley & Worth Valley Railway   OP

UNCLASSIFIED   0-4-0 DH

Built: 1960 by North British Locomotive Company, Glasgow.
Engine: MAN W6V 17.5/22A of 168 kW at 1100 rpm.
Length over buffers: 7.455 m.   **Weight:** 36.00 t.
Extreme width: **Wheel Diameter:** 1143 mm.
Extreme height: **Maximum Permitted Speed:** 15 mph.
D2767   06.67   East Lancashire Railway   UR
D2774   06.67   East Lancashire Railway   UR

UNCLASSIFIED   Bo-Bo

Built: 1903 by North Eastern Railway, Gateshead Works.
Supply System: 600-630 V dc third rail or overhead.
Traction Motors: Four BTH design.
Length over buffers: 11.506 m.   **Weight:** 46.00 t.
Extreme width: **Wheel Diameter:** 914 mm.
Extreme height: **Maximum Permitted Speed:**
Original Number: Originally NER 1, then LNER 6480 and BR 26500.
E26500   02.64   National Railway Museum   CE

UNCLASSIFIED   Bo

Built: 1898 by Siemens, London.
Supply System: 500 v dc inner live rail.
Traction Motors: Two Siemens of 45 kW.
Length over buffers: **Weight:** 22.00 t.
Extreme width: **Wheel Diameter:** 1016 mm.
Extreme height: **Maximum Permitted Speed:**
Original Number: Originally un-numbered, then Southern Railway 75s.
DS75   05.68   National Railway Museum   CE

UNCLASSIFIED   Bo

Built: 1917 by North Staffordshire Railway, Stoke-on-Trent Works.
Supply System: 108 cell battery.
Traction Motors: Two BTH of 31 kW each.

Length over buffers:		**Weight:** 17.00 t.	
Extreme width:		**Wheel Diameter:** 940 mm.	
Extreme height:		**Maximum Permitted Speed:**	

Original Number: Originally numbered NSR 1.

BEL 2	03.63	National Railway Museum	CE

UNCLASSIFIED   0-4-0 DM

Built: 1958 by Ruston & Hornsby, Lincoln.
Gauge: 914 mm. (For Beeston Sleeper Depot).
Engine: Ruston 4YCL of 36 kW at 1375 rpm.

Length over buffers: 4.140 m.	**Weight:** 8.20 t.
Extreme width:	**Wheel Diameter:** 762 mm.
Extreme height:	**Maximum Permitted Speed:**

ED10	02.65	Irchester Narrow Gauge Railway Museum	UR

UNCLASSIFIED   0-4-0 DM

Built: 1957 by Ruston & Hornsby, Lincoln.
Gauge: 457 mm. (For BR, Horwich Works).
Engine: Ruston 2VSH of 15 kW at 1200 rpm.

Length over buffers:	**Weight:** 3.50 t.
Extreme width:	**Wheel Diameter:** 414 mm.
Extreme height:	**Maximum Permitted Speed:**

ZM32	03.64	Narrow Gauge Railway Centre	CE

UNCLASSIFIED   0-4-0 DM

Built: 1956-57 by Ruston & Hornsby, Lincoln.
Gauge: 610 mm. (For Chesterton Junction Central Materials Depot).
Engine: Ruston 3VSHL of 24 kW.

Length over buffers: 2.261 m.	**Weight:** 4.00 t.
Extreme width:	**Wheel Diameter:** 414 mm.
Extreme height:	**Maximum Permitted Speed:**

85049	07.86	Avon & Radstock Railway	OP
85051	07.86	Cadeby Light Railway	OP

UNCLASSIFIED   0-4-0 DM

Built: 1919 by Motor Rail & Tram Car Company, Bedford, for the Lancashire & Yorkshire Railway.
Engine: Dorman 4JO of 30 kW (Petrol).

Length over buffers: 4.064 m.	**Weight:** 8.00 t.	
Extreme width:	**Wheel Diameter:** 940 mm.	
Extreme height:	**Maximum Permitted Speed:** 7 mph.	

| 1 | 11.30 | Chasewater Light Railway | UR |

UNCLASSIFIED   0-4-0 DM

Built: 1934 by English Electric, Preston, for the London Midland & Scottish Railway.
Engine: Allan 8RS18 of 119 kW at 1200 rpm.

Length over buffers: 7.277 m.	**Weight:** 25.40 t.	
Extreme width:	**Wheel Diameter:** 914 mm.	
Extreme height:	**Maximum Permitted Speed:** 12 mph.	

Subsequent Number: War Department 240.

| 7050 | 03.43 | Museum of Army Transport | CE |

UNCLASSIFIED   0-6-0 DM

Built: 1932 by Hunslet Engine Company, Leeds, for the London Midland & Scottish Railway.
Engine: Ricardo-MAN 6 cylinder of 112 kW at 900 rpm.

Length over buffers: 7.061 m.	**Weight:** 21.40 t.	
Extreme width:	**Wheel Diameter:** 914 mm.	
Extreme height:	**Maximum Permitted Speed:** 30 mph.	

Subsequent Number: 7401.

| 7051 | 12.45 | Middleton Railway | SU |

UNCLASSIFIED   0-6-0 DE

Built: 1935 by Hawthorn Leslie, Newcastle-upon-Tyne, for the London Midland & Scottish Railway.

33

Engine: English Electric 6KT of 261 kW at 675 rpm.
Length over buffers: 8.706 m.   **Weight:** 51.45 t.
Extreme width: **Wheel Diameter:** 1232 mm.
Extreme height: **Maximum Permitted Speed:** 30 mph.
Subsequent Number: War Department 18.

7069	12.40	East Lancashire Railway	UR

UNCLASSIFIED   0-6-0 DE

Built: 1941 by London Midland & Scottish Railway, Derby Locomotive Works.
Engine: English Electric 6KT of 261 kW at 680 rpm.
Length over buffers: 9.893 m.   **Weight:** 53.50 t.
Extreme width: **Wheel Diameter:** 1295 mm.
Extreme height: **Maximum Permitted Speed:** 20 mph.
Subsequent Numbers: WD 52, FS 700.001 (7103); WD 55, FS 700.003 (7106).

7103	12.42	FSAS, Arezzo, Italy	OP
7106	12.42	Carbiboni, Colico, Italy	OP

1.4 FORMER WAR DEPARTMENT LOCOMOTIVES

Note: Dates shown in this section are dates withdrawn from War Department/ Ministry of Defence service.

UNCLASSIFIED   0-6-0 DE

Built: 1945 by London Midland & Scottish Railway, Derby Locomotive Works.
Engine: English Electric 6KT of 261 kW at 680 rpm.
Length over buffers: 8.877 m.   **Weight:** 47.25 t.
Extreme width: **Wheel Diameter:** 1232 mm.
Extreme height: **Maximum Permitted Speed:** 20 mph.
Subsequent Numbers: NS 508 (70269).

70269	03.46	Rotterdam Maritime Museum, The Netherlands	CE
70272	??.80	Lakeside & Haverthwaite Railway	OP

1.5 MANUFACTURERS PROTOTYPES NOT TAKEN INTO BR STOCK

UNCLASSIFIED   Co-Co DE

Built: 1954 by English Electric, Preston.
Engines: Two Napier Deltic D18-25 of 1230 kW each at 1500 rpm.
Length over buffers: 20.117 m.   **Weight:** 106.00 t.
Extreme width: 2.680 m.   **Wheel Diameter:** 1092 mm.
Extreme height: 3.924 m.   **Maximum Permitted Speed:** 90 mph.
DELTIC   03.61   National Railway Museum   IE

UNCLASSIFIED   0-6-0 DE

Built: 1956 by English Electric, Vulcan Foundry.
Engine: English Electric 6RKT of 373 kW at 750 rpm.
Length over buffers: **Weight:** 48.00 t.
Extreme width: **Wheel Diameter:** 1219 mm.
Extreme height: **Maximum Permitted Speed:** 35 mph.
Original Number: D226.
D0226   10.60   Keighley & Worth Valley Railway   OP

UNCLASSIFIED   0-4-0 DH

Built: 1954 by North British Locomotive Company, Glasgow.
Engine: Paxman 6VRPHXL of 233 kW at 1250 rpm.
Length over buffers: **Weight:**
Extreme width: **Wheel Diameter:** 1016 mm.
Extreme height: **Maximum Permitted Speed:** 12 mph.

TOM   ??.55   Telford Horsehay Steam Trust   UR
TIGER   ??.55   Bo'ness & Kinneil Railway   OP

CLASS 103 — DMBS

Built: 1958 by Park Royal Vehicles, London.
Engines: Two horizontal AEC 220 type of 112 kW at 1800 rpm.
Transmission: Mechanical.

Length over buffers: 18.491 m.		**Weight:** 33.45 t.	
Extreme width: 2.724 m.		**Seats:** 52.	
Extreme height: 3.766 m.		**Toilet:** Not equipped.	

Maximum Permitted Speed: 70 mph.

| 50397 | 02.71 | Battlefield Steam Railway | SU |
| 50413 | 12.72 | West Somerset Railway | UR |

UNCLASSIFIED — DMBS

Built: 1957-58 by D. Wickham, Ware.
Engines: Two horizontal Leyland 680/1 type of 112 kW at 1800 rpm.
Transmission: Mechanical.

Length over buffers: 18.339 m.		**Weight:** 27.00 t.	
Extreme width: 2.819 m.		**Seats:** 59.	
Extreme height: 3.835 m.		**Toilet:** Not equipped.	

Maximum Permitted Speed: 70 mph.

50415	09.61	Trinidad Government Railways	SU
50416	10.67	Midland Railway Centre	UR
50419	09.61	Trinidad Government Railways	SU

CLASS 126 — DMSL

Built: 1959 by BR at Swindon Works.
Engines: Two horizontal AEC 220 type of 112 kW at 1800 rpm.
Transmission: Mechanical.

Length over buffers: 20.155 m.		**Weight:** 38.85 t.	
Extreme width: 2.819 m.		**Seats:** 64.	
Extreme height: 3.886 m.		**Toilet:** 1.	

Maximum Permitted Speed: 70 mph.

| 51017 | 12.82 | Bo'ness & Kinneil Railway | UR |

CLASS 126 | DMBSL

Built: 1959 by BR at Swindon Works.
Engines: Two horizontal AEC 220 type of 112 kW at 1800 rpm.
Transmission: Mechanical.

Length over buffers: 20.155 m.	**Weight:** 37.35 t.
Extreme width: 2.819 m.	**Seats:** 52.
Extreme height: 3.886 m.	**Toilets:** 2.

Maximum Permitted Speed: 70 mph.

51043	12.82	Bo'ness & Kinneil Railway	UR

CLASS 100 | DMBS

Built: 1957 by Gloucester Railway Carriage & Wagon Company, Gloucester
Engines: Two horizontal AEC 220 type of 112 kW at 1800 rpm.
Transmission: Mechanical.

Length over buffers: 18.491 m.	**Weight:** 30.00 t.
Extreme width: 2.743 m.	**Seats:** 52.
Extreme height: 3.861 m.	**Toilet:** Not equipped.

Maximum Permitted Speed: 70 mph.

51118	10.72	West Somerset Railway	SS

CLASS 116 | DMBS

Built: 1958 by BR at Derby C & W Works.
Engines: Two horizontal Leyland type 680/1 of 112 kW at 1800 rpm.
Transmission: Mechanical.

Length over buffers: 20.447 m.	**Weight:** 35.95 t.
Extreme width: 2.819 m.	**Seats:** 65.
Extreme height: 3.861 m.	**Toilet:** Not equipped.

Maximum Permitted Speed: 70 mph.

51131	09.93	Battlefield Line	OP
51134	05.93	PD Fuels, Coed Bach Disposal Point	
51135	05.92	Swansea Vale Railway	OP

CLASS 116 | DMS

Built: 1958 by BR at Derby C & W Works.
Engines: Two horizontal Leyland type 680/1 of 112 kW at 1800 rpm.
Transmission: Mechanical.

Length over buffers: 20.447 m.	**Weight:** 35.95 t.
Extreme width: 2.819 m.	**Seats:** 95.

Extreme height: 3.861 m. **Toilet:** Not equipped.
Maximum Permitted Speed: 70 mph.

51147	05.93	PD Fuels, Coed Bach Disposal Point	
51148	06.92	Swansea Vale Railway	OP

CLASS 101 DMBS

Built: 1959 by Metropolitan Cammell, Birmingham.
Engines: Two horizontal AEC 220 type of 112 kW at 1800 rpm.
Transmission: Mechanical.
Length over buffers: 18.491 m. **Weight:** 32.05 t.
Extreme width: 2.819 m. **Seats:** 52.
Extreme height: 3.848 m. **Toilet:** Not equipped.
Maximum Permitted Speed: 70 mph.

51203	09.89	Darlington Railway Museum	SU

CLASS 117 DMBS

Built: 1959-60 by Pressed Steel, Linwood.
Engines: Two horizontal Leyland type 680/1 of 112 kW at 1800 rpm.
Transmission: Mechanical.
Length over buffers: 20.447 m. **Weight:** 36.50 t.
Extreme width: 2.819 m. **Seats:** 65.
Extreme height: 3.861 m. **Toilet:** Not equipped.
Maximum Permitted Speed: 70 mph.

51342	12.93	Lavender Line	OP
51346	07.93	North Norfolk Railway	OP
51347	12.93	Mangapps Farm Railway	OP
51351	12.93	Chinnor & Princes Risborough Railway	UR
51360	09.93	East Lancashire Railway	OP
51364	03.94	Emerson Railway Services	SB
51367	02.93	Northampton & Lamport Railway	OP
51370	01.94	Chasewater Railway	UR
51372	01.94	Chasewater Railway	UR

CLASS 117 DMS

Built: 1959-60 by Pressed Steel, Linwood.
Engines: Two horizontal Leyland type 680/1 of 112 kW at 1800 rpm.
Transmission: Mechanical.
Length over buffers: 20.447 m. **Weight:** 36.50 t.
Extreme width: 2.819 m. **Seats:** 89.
Extreme height: 3.861 m. **Toilet:** Not equipped.
Maximum Permitted Speed: 70 mph.

51384	12.93	Lavender Line	OP
51388	07.93	North Norfolk Railway	OP
51397	12.93	Chinnor & Princes Risborough Railway	UR
51402	09.93	East Lancashire Railway	OP
51406	01.94	Emerson Railway Services	SB
51409	01.93	Emerson Railway Services	SB
51412	01.94	Chasewater Railway	UR

CLASS 105   DMBS

Built: 1959 by Cravens, Sheffield.
Engines: Two horizontal AEC 220 type of 112 kW at 1800 rpm.
Transmission: Mechanical.
Length over buffers: 18.644 m.   **Weight:** 30.00 t.
Extreme width: 2.819 m.   **Seats:** 52.
Extreme height: 3.772 m.   **Toilet:** Not equipped.
Maximum Permitted Speed: 70 mph.

| 51485 | 05.81 | West Somerset Railway | OP |

CLASS 108   DMCL

Built: 1959-60 by BR at Derby C & W Works.
Engines: Two horizontal Leyland 680/1 or 680/13 type of 112 kW at 1800 rpm.
Transmission: Mechanical.
Length over buffers: 18.491 m.   **Weight:** 28.50 t.
Extreme width: 2.819 m.   **Seats:** 12 1st, 52 2nd.
Extreme height: 3.874 m.   **Toilet:** 1.
Maximum Permitted Speed: 70 mph.

51562	06.92	National Railway Museum	UR
51565	03.92	Keighley & Worth Valley Railway	OP
51566	12.92	Peak Rail	OP
51568	02.93	East Anglian Railway Museum	OP
51571	09.92	Kent & East Sussex Steam Railway	OP
51572	01.93	East Kent Light Railway	OP

CLASS 127   DMBS

Built: 1959 by BR at Derby C & W Works.
Engines: Two Rolls Royce C8NFLH823 type of 178 kW at 1800 rpm.
Transmission: Hydraulic.
Length over buffers: 20.447 m.   **Weight:** 39.85 t.
Extreme width: 2.819 m.   **Seats:** 76.
Extreme height: 3.874 m.   **Toilet:** Not equipped.
Maximum Permitted Speed: 70 mph.

51592	01.84	South Devon Railway	OP
51604	08.83	South Devon Railway	OP
51616	01.84	Great Central Railway	OP
51618	12.83	Llangollen Railway	OP
51622	01.84	Great Central Railway	OP

CLASS 115   DMBS

Built: 1960 by BR at Derby C & W Works.
Engines: Two horizontal Leyland Albion 902 type of 172 kW at 1800 rpm.
Transmission: Mechanical.
Length over buffers: 20.447 m.   **Weight:** 37.85 t.
Extreme width: 2.819 m.   **Seats:** 78.
Extreme height: 3.874 m.   **Toilet:** Not equipped.
Maximum Permitted Speed: 70 mph.

51655	08.92	Lavender Line	OP
51663	08.92	West Somerset Railway	OP
51669	03.92	Mid-Norfolk Railway	OP
51677	08.92	Lavender Line	OP

CLASS 110   DMBC

Built: 1961 by Birmingham Railway Carriage & Wagon Company, Smethwick.
Engines: Two Rolls Royce C6NFLH138D type of 134 kW at 1800 rpm.
Transmission: Mechanical.
Length over buffers: 18.485 m.   **Weight:** 32.00 t.
Extreme width: 2.819 m.   **Seats:** 12 1st, 33 2nd.
Extreme height: 3.835 m.   **Toilet:** Not equipped.
Maximum Permitted Speed: 70 mph.

| 51813 | 03.90 | East Lancashire Railway | OP |

CLASS 110   DMCL

Built: 1961 by Birmingham Railway Carriage & Wagon Company, Smethwick.
Engines: Two Rolls Royce C6NFLH138D type of 134 kW at 1800 rpm.
Transmission: Mechanical.
Length over buffers: 18.485 m.   **Weight:** 32.00 t.
Extreme width: 2.819 m.   **Seats:** 12 1st, 54 2nd.
Extreme height: 3.835 m.   **Toilet:** 1.
Maximum Permitted Speed: 70 mph.

| 51842 | 02.90 | East Lancashire Railway | OP |

CLASS 115   DMBS

Built: 1960 by BR at Derby C & W Works.
Engines: Two horizontal Leyland Albion 902 type of 172 kW at 1800 rpm.
Transmission: Mechanical.
Length over buffers: 20.447 m.   **Weight:** 37.85 t.
Extreme width: 2.819 m.   **Seats:** 78.
Extreme height: 3.874 m.   **Toilet:** Not equipped.
Maximum Permitted Speed: 70 mph.

51849	02.92	Mid-Norfolk Railway	UR
51852	09.93	West Somerset Railway	UR
51886	08.92	Buckinghamshire Railway Centre	UR
51887	08.92	West Somerset Railway	UR
51899	08.92	Buckinghamshire Railway Centre	UR

CLASS 108   DMBS

Built: 1960-61 by BR at Derby C & W Works.
Engines: Two horizontal Leyland 680/1 or 680/13 type of 112 kW at 1800 rpm.
Transmission: Mechanical.
Length over buffers: 18.491 m.   **Weight:** 29.05 t.
Extreme width: 2.819 m.   **Seats:** 52.
Extreme height: 3.874 m.   **Toilet:** Not equipped.
Maximum Permitted Speed: 70 mph.

51907	02.93	Llangollen Railway	OP
51909	07.93	MOD Long Marston (stored for Avon Valley)	SU
51914	04.93	Dean Forest Railway	UR
51919	02.93	Swanage Railway	UR
51922	06.92	National Railway Museum	UR
51933	02.93	Peak Rail	SU
51935	10.92	Severn Valley Railway	SU
51937	09.92	Peak Rail	UR
51941	11.90	Severn Valley Railway	OP
51942	06.93	Pontypool & Blaenavon Railway	UR
51947	02.93	Bodmin & Wenford Railway	SU
51950	06.91	Gloucestershire Warwickshire Railway	OP

CLASS 107   DMBS

Built: 1961 by BR at Derby C & W Works.
Engines: Two horizontal Leyland 1595 type of 112 kW at 1800 rpm.
Transmission: Mechanical.
Length over buffers: 18.491 m.   **Weight:** 34.45 t.

| Extreme width: 2.819 m. | Seats: 52. |
| Extreme height: 3.874 m. | Toilet: Not equipped. |

Maximum Permitted Speed: 70 mph.

| 52006 | 10.92 | East Kent Light Railway | OP |
| 52008 | 10.92 | Strathspey Railway | OP |

CLASS 107   DMCL

Built: 1961 by BR at Derby C & W Works.
Engines: Two horizontal Leyland 1595 type of 112 kW at 1800 rpm.
Transmission: Mechanical.

Length over buffers: 18.491 m.	Weight: 34.45 t.
Extreme width: 2.819 m.	Seats: 12 1st, 53 2nd.
Extreme height: 3.874 m.	Toilet: 1.

Maximum Permitted Speed: 70 mph.

| 52029 | 10.92 | Lakeside & Haverthwaite Railway | SU |
| 52031 | 10.92 | East Kent Light Railway | OP |

CLASS 108   DMCL

Built: 1960-61 by BR at Derby C & W Works.
Engines: Two horizontal Leyland 680/1 type of 112 kW at 1800 rpm.
Transmission: Mechanical.

Length over buffers: 18.491 m.	Weight: 28.05 t.
Extreme width: 2.794 m.	Seats: 12 1st, 50 2nd.
Extreme height: 3.874 m.	Toilet: 1.

Maximum Permitted Speed: 70 mph.

52044	02.93	Pontypool & Blaenarvon Railway	UR
52048	02.93	Swanage Railway	UR
52053	09.92	Storage at Horsham (for East Anglian RM)	SU
52054	02.93	Bodmin & Wenford Railway	SU
52062	06.91	Gloucestershire Warwickshire Railway	OP
52064	11.90	Severn Valley Railway	OP

CLASS 110   DMBC

Built: 1962 by Birmingham Railway Carriage & Wagon Company, Smethwick.
Engines: Two Rolls Royce C6NFLH138D type of 134 kW at 1800 rpm.
Transmission: Mechanical.

Length over buffers: 18.485 m.	Weight: 32.00 t.
Extreme width: 2.819 m.	Seats: 12 1st, 33 2nd.
Extreme height: 3.835 m.	Toilet: Not equipped.

Maximum Permitted Speed: 70 mph.

| 52071 | 03.90 | Lakeside & Haverthwaite Railway | OP |

CLASS 110 — DMCL

Built: 1961 by Birmingham Railway Carriage & Wagon Company, Smethwick.
Engines: Two Rolls Royce C6NFLH138D type of 134 kW at 1800 rpm.
Transmission: Mechanical.
Length over buffers: 18.485 m.   **Weight:** 32.00 t.
Extreme width: 2.819 m.   **Seats:** 12 1st, 54 2nd.
Extreme height: 3.835 m.   **Toilet:** 1.
Maximum Permitted Speed: 70 mph.

52077		03.90	Lakeside & Haverthwaite Railway	OP

CLASS 114 — DMBS

Built: 1956 by BR at Derby C & W Works.
Engines: Two horizontal Leyland type 680/1 of 112 kW at 1800 rpm.
Transmission: Mechanical.
Length over buffers: 20.447 m.   **Weight:** 37.50 t.
Extreme width: 2.819 m.   **Seats:** 62.
Extreme height: 3.861 m.   **Toilet:** Not equipped.
Maximum Permitted Speed: 70 mph.

53019	50019	06.92	Midland Railway Centre	OP

CLASS 104 — DMBS

Built: 1957-58 by Birmingham Railway Carriage & Wagon Company, Smethwick.
Engines: Two horizontal Leyland 680/1 type of 112 kW (150 hp).
Transmission: Mechanical.
Length over buffers: 18.491 m.   **Weight:** 31.00 t.
Extreme width: 2.819 m.   **Seats:** 52.
Extreme height: 3.835 m.   **Toilet:** Not equipped.
Maximum Permitted Speed: 70 mph.

53437	50437	02.92	BNFL, Salwick	SU
53447	50447	03.92	Llangollen Railway	SP
53454	50454	03.92	Llangollen Railway	OP
53455	50455	09.92	Cheddleton Railway Centre	OP
53479	50479	02.92	Cambrian Railways Society	UR

CLASS 104 — DMCL

Built: 1957-58 by Birmingham Railway Carriage & Wagon Company, Smethwick.
Engines: Two horizontal Leyland 680/1 type of 112 kW (150 hp).
Transmission: Mechanical.

Length over buffers: 18.491 m.   **Weight:** 31.00 t.
Extreme width: 2.819 m.   **Seats:** 12 1st, 54 (* 51) 2nd.
Extreme height: 3.835 (* 3.880) m.   **Toilet:** 1.
Maximum Permitted Speed: 70 mph.

53494	50494	06.90	Cheddleton Railway Centre	SB
53517	50517	05.90	Cheddleton Railway Centre	UR
53528	50528	03.92	Llangollen Railway	OP
53531	50531	03.92	Cambrian Railways Society	UR
53556*	50556	05.89	South Yorkshire Railway	SU

CLASS 108   DMBS

Built: 1958 by BR at Derby C & W Works.
Engines: Two horizontal Leyland 680/1 type of 112 kW (150 hp).
Transmission: Mechanical.
Length over buffers: 18.491 m.   **Weight:** 29.05 t.
Extreme width: 2.794 m.   **Seats:** 52.
Extreme height: 3.874 m.   **Toilet:** Not equipped.
Maximum Permitted Speed: 70 mph.

53599	50599	01.93	East Anglian Railway Museum	OP
53619	50619	07.91	Dean Forest Railway	OP

CLASS 108   DMCL

Built: 1958 by BR at Derby C & W Works.
Engines: Two horizontal Leyland 680/1 type of 112 kW at 1800 rpm.
Transmission: Mechanical.
Length over buffers: 18.491 m.   **Weight:** 28.05 t.
Extreme width: 2.794 m.   **Seats:** 12 1st, 50 2nd.
Extreme height: 3.874 m.   **Toilet:** 1.
Maximum Permitted Speed: 70 mph.

53628	50628	10.93	Southall Railway Centre	UR
53632	50632	02.93	Pontypool & Blaenarvon Railway	UR
53645	50645	02.93	Bodmin & Wenford Railway	SU

CLASS 108   DMBS

Built: 1959-60 by BR at Derby C & W Works.
Engines: Two horizontal Leyland 680/1 type of 112 kW (150 hp).
Transmission: Mechanical.
Length over buffers: 18.491 m.   **Weight:** 29.05 t.
Extreme width: 2.794 m.   **Seats:** 52.
Extreme height: 3.874 m.   **Toilet:** Not equipped.

Maximum Permitted Speed: 70 mph.

53928	50928	03.92	Keighley & Worth Valley Railway	OP
53933	50933	11.92	Peak Rail	OP
53971	50971	02.93	Kent & East Sussex Steam Railway	OP
53980	50980	02.93	Bodmin & Wenford Railway	OP

CLASS 114 DTCL

Built: 1956-57 by BR at Derby C & W Works.
Length over buffers: 20.447 m.   **Weight:** 29.55 t.
Extreme width: 2.819 m.   **Seats:** 12 1st, 62 2nd.
Extreme height: 3.874 m.   **Toilet:** 1.
Maximum Permitted Speed: 70 mph.

| 54006 | 56006 | 03.92 | Midland Railway Centre | OP |
| 54047 | 56047 | 10.91 | Strathspey Railway | OP |

CLASS 108 DTCL

Built: 1958-60 by BR at Derby C & W Works.
Length over buffers: 18.491 m.   **Weight:** 21.15 t.
Extreme width: 2.737 m.   **Seats:** 12 1st, 53 2nd.
Extreme height: 3.874 m.   **Toilet:** 1.
Maximum Permitted Speed: 70 mph.

54207	56207	07.90	BSC, Scunthorpe	OH
54208	56208	03.93	Caerphilly Railway Society	OH
54223	56223	06.93	Southall Railway Centre	UR
54224	56224	09.92	East Kent Light Railway	OP
54270	56270	07.93	Pontypool & Blaenarvon Railway	UR
54271	56271	07.93	MOD Long Marston (stored for Avon Valley)	SU
54274	56274	02.92	Rutland Railway Museum	OH
54279	56279	08.93	Wales Railway Centre	OH

CLASS 121 DTS

Built: 1961 by Pressed Steel, Linwood.
Length over buffers: 20.447 m.   **Weight:** 29.35 t.
Extreme width: 2.819 m.   **Seats:** 91.
Extreme height: 3.861 m.   **Toilet:** Not equipped.
Maximum Permitted Speed: 70 mph.

| 54287 | 56287 | 04.92 | Mangapps Farm Railway Museum | OP |
| 54289 | 56289 | 12.92 | Battlefield Steam Railway | OP |

CLASS 105 DTCL

Built: 1959 by Cravens, Sheffield.
Length over buffers: 18.491 m.   **Weight:** 24.10 t.
Extreme width: 2.819 m.   **Seats:** 12 1st, 51 2nd.
Extreme height: 3.772 m.   **Toilet:** 1.
Maximum Permitted Speed: 70 mph.

54456	56456	07.83	Llangollen Railway	OP

CLASS 108 DTCL

Built: 1960 by BR at Derby C & W Works.
Length over buffers: 18.491 m.   **Weight:** 21.15 t.
Extreme width: 2.794 m.   **Seats:** 12 1st, 53 2nd.
Extreme height: 3.874 m.   **Toilet:** 1.
Maximum Permitted Speed: 70 mph.

54484	56484	07.92	Peak Rail	UR
54490	56490	02.93	Llangollen Railway	OP
54491	56491	12.92	East Anglian Railway Museum	OP
54492	56492	07.91	Dean Forest Railway	OP
54495	56495	10.93	Northampton & Lamport Railway	OP
54504	56504	02.93	Peak Rail	SU

CLASS 122 DMBS

Built: 1958 by Gloucester Railway Carriage & Wagon Company, Gloucester.
Engines: Two horizontal AEC 220 type of 112 kW at 1800 rpm.
Transmission: Mechanical.
Length over buffers: 20.447 m.   **Weight:** 36.50 t.
Extreme width: 2.743 m.   **Seats:** 65.
Extreme height: 3.861 m.   **Toilet:** Not equipped.
Maximum Permitted Speed: 70 mph.
Subsequent Number: 55032 DB 977842 in departmental service

55000	01.94	P. Waterman	SB
55005	10.92	Battlefield Steam Railway	OP
55032	12.92	East Lancashire Railway	OP

CLASS 127 DMPMV

Built: 1959 by BR at Derby C & W Works. Converted for parcels use 1985.
Engines: Two Rolls Royce C8NFLH823 type of 178 kW at 1800 rpm.
Transmission: Hydraulic.

Length over buffers: 20.447 m.   **Weight:** 39.85 t.
Extreme width: 2.819 m.   **Carrying Capacity:** 10 t.
Extreme height: 3.874 m.   **Toilet:** Not equipped.
Maximum Permitted Speed: 70 mph.

55966	51591	05.89	Midland Railway Centre	UR
55967	51610	05.89	Rogart Station, Highland	CC
55976	51625	05.89	Midland Railway Centre	UR

CLASS 100   DTCL

Built: 1957 by Gloucester Railway Carriage & Wagon Company, Gloucester.
Length over buffers: 18.491 m.   **Weight:** 25.10 t.
Extreme width: 2.819 m.   **Seats:** 12 1st, 54 2nd.
Extreme height: 3.874 m.   **Toilet:** 1.
Maximum Permitted Speed: 70 mph.

| 56097 | 10.72 | West Somerset Railway | SS |

CLASS 105   DTCL

Built: 1957 by Cravens, Sheffield.
Length over buffers: 18.491 m.   **Weight:** 23.15 t.
Extreme width: 2.819 m.   **Seats:** 12 1st, 51 2nd.
Extreme height: 3.835 m.   **Toilet:** 1.
Maximum Permitted Speed: 70 mph.

| 56121 | 05.81 | West Somerset Railway | OP |

CLASS 103   DTCL

Built: 1958 by Park Royal Vehicles, London.
Length over buffers: 18.491 m.   **Weight:** 26.55 t.
Extreme width: 2.819 m.   **Seats:** 16 1st, 48 2nd.
Extreme height: 3.772 m.   **Toilet:** 1.
Maximum Permitted Speed: 70 mph.

| 56160 | 02.71 | Battlefield Steam Railway | SU |
| 56169 | 12.72 | West Somerset Railway | UR |

UNCLASSIFIED   DTCL

Built: 1957-58 by D. Wickham, Ware.
Length over buffers: 18.491 m.   **Weight:** 22.50 t.
Extreme width: 2.819 m.   **Seats:** 16 1st, 50 2nd.
Extreme height: 3.835 m.   **Toilet:** 1.

56170	09.61	Trinidad Government Railways	SU
56171	10.67	Midland Railway Centre	UR
56174	09.61	Trinidad Government Railways	SU

CLASS 100   DTCL

Built: 1957-58 by Gloucester Railway Carriage & Wagon Company, Gloucester.
Length over buffers: 18.491 m.   **Weight:** 25.10 t.
Extreme width: 2.819 m.   **Seats:** 12 1st, 54 2nd.
Extreme height: 3.874 m.   **Toilet:** 1.
Maximum Permitted Speed: 70 mph.

| 56301 | 02.72 | BNFL, Salwick | SU |
| 56317 | 04.74 | Gwili Railway/Rheilffordd Gwili | SU |

CLASS 116   TC

Built: 1957 by BR at Derby C & W Works.
Length over buffers: 20.447 m.   **Weight:** 28.55 t.
Extreme width: 2.819 m.   **Seats:** 28 1st, 74 2nd.
Extreme height: 3.874 m.   **Toilet:** Not equipped.
Maximum Permitted Speed: 70 mph.

| 59003 | 11.83 | Paignton & Dartmouth Steam Railway | OH |
| 59004 | 11.83 | Paighton & Dartmouth Steam Railway | OH |

CLASS 104   TCL

Built: 1957 by Birmingham Railway Carriage & Wagon Company, Smethwick.
Length over buffers: 18.491 m.   **Weight:** 24.10 t.
Extreme width: 2.819 m.   **Seats:** 12 1st, 54 2nd.
Extreme height: 3.880 m.   **Toilet:** 1.
Maximum Permitted Speed: 70 mph.

| 59137 | 10.89 | Cheddleton Railway Centre | SB |

CLASS 104   TBSL

Built: 1958 by Birmingham Railway Carriage & Wagon Company, Smethwick.
Length over buffers: 18.491 m.   **Weight:** 25.00 t.
Extreme width: 2.819 m.   **Seats:** 51.
Extreme height: 3.880 m.   **Toilet:** 1.
Maximum Permitted Speed: 70 mph.

| 59228 | 01.91 | | SB |

CLASS 108   TBSL

Built: 1958 by BR at Derby C & W Works.
Length over buffers: 18.491 m.   **Weight:** 23.15 t.
Extreme width: 2.794 m.   **Seats:** 50.
Extreme height: 3.874 m.   **Toilet:** 1.
Maximum Permitted Speed: 70 mph.

59245	07.90	BSC, Scunthorpe	OH
59250	07.91	Severn Valley Railway	OP

CLASS 120   TSLRB

Built: 1958 BR at Swindon Works.
Length over buffers: 20.447 m.   **Weight:** 30.50 t.
Extreme width: 2.819 m.   **Seats:** 60.
Extreme height: 3.899 m.   **Toilet:** 1.
Maximum Permitted Speed: 70 mph.

59276	11.83	Great Central Railway	OP

CLASS 108   TSL

Built: 1958 by BR at Derby C & W Works.
Length over buffers: 18.491 m.   **Weight:** 23.50 t.
Extreme width: 2.794 m.   **Seats:** 68.
Extreme height: 3.874 m.   **Toilet:** 1.
Maximum Permitted Speed: 70 mph.

59387	11.92	Peak Rail	UR

CLASS 126   TCL

Built: 1959 by BR at Swindon Works.
Length over buffers: 20.155 m.   **Weight:** 31.50 t.
Extreme width: 2.819 m.   **Seats:** 18 1st, 32 2nd.
Extreme height: 3.899 m.   **Toilets:** 2.
Maximum Permitted Speed: 70 mph.

59404	12.82	Bo'ness & Kinneil Railway	UR

CLASS 116   TC

Built: 1958 by BR at Derby C & W Works.
Length over buffers: 20.447 m.   **Weight:** 28.55 t.
Extreme width: 2.819 m.   **Seats:** 28 1st, 74 2nd.
Extreme height: 3.874 m.   **Toilet:** Not equipped.

Maximum Permitted Speed: 70 mph.

59444	07.90	Chasewater Railway	OH
59445	10.92	Swansea Vale Railway	OP

CLASS 117 TCL

Built: 1959-60 by Pressed Steel, Linwood.
Length over buffers: 20.447 m. **Weight:** 30.50 t.
Extreme width: 2.819 m. **Seats:** 22 1st, 48 2nd.
Extreme height: 3.861 m. **Toilets:** 2.
Maximum Permitted Speed: 70 mph.

59488	10.93	Paignton & Dartmouth Steam Railway	SU
59490	01.94	East Lancashire Railway	OP
59494	07.93	Paignton & Dartmouth Steam Railway	SU
59496	03.94	Battlefield Steam Railway	UR
59501	03.94	Chinnor & Princes Risborough Railway	UR
59503	01.93	Paignton & Dartmouth Steam Railway	UR
59507	03.94	Paignton & Dartmouth Steam Staem	OH
59508	04.93	Battlefield Steam Railway	UR
59511	10.93	Lavender Line	UR
59513	03.94	Paignton & Dartmouth Steam Railway	UR
59514	07.93	West Somerset Railway	UR
59516	01.94	North Norfolk Railway	OP
59517	09.93	Paignton & Dartmouth Steam Railway	OH
59522	09.93	Battlefield Steam Railway	OP

CLASS 111 TSLRB

Built: 1960 by Metropolitan Cammell, Birmingham.
Length over buffers: 18.491 m. **Weight:** 25.10 t.
Extreme width: 2.819 m. **Seats:** 53.
Extreme height: 3.848 m. **Toilet:** 1.
Maximum Permitted Speed: 70 mph.

59575	09.75	Museum of Science & Industry, Manchester	IE

CLASS 127 TSL

Built: 1959 by BR at Derby C & W Works.
Length over buffers: 20.447 m. **Weight:** 30.00 t.
Extreme width: 2.819 m. **Seats:** 86.
Extreme height: 3.861 m. **Toilets:** 2.
Maximum Permitted Speed: 70 mph.

59603	10.93	Chasewater Railway	UR
59609	09.93	Midland Railway Centre	SU

CLASS 115   TS

Built: 1960 by BR at Derby C & W Works.
Length over buffers: 20.447 m.   **Weight:** 29.05 t.
Extreme width: 2.819 m.   **Seats:** 106.
Extreme height: 3.861 m.   **Toilet:** Not equipped.
Maximum Permitted Speed: 70 mph.

| 59659 | 08.92 | South Devon Railway | OH |

CLASS 115   TCL

Built: 1960 by BR at Derby C & W Works.
Length over buffers: 20.447 m.   **Weight:** 30.00 t.
Extreme width: 2.819 m.   **Seats:** 30 1st, 40 2nd.
Extreme height: 3.861 m.   **Toilet:** 1.
Maximum Permitted Speed: 70 mph.

| 59664 | 08.92 | Mid-Norfolk Railway | UR |
| 59678 | 08.92 | West Somerset Railway | UR |

CLASS 110   TSL

Built: 1961 by Birmingham Railway Carriage & Wagon Company, Smethwick.
Length over buffers: 18.478 m.   **Weight:** 24.10 t.
Extreme width: 2.819 m.   **Seats:** 72.
Extreme height: 3.874 m.   **Toilet:** 1.
Maximum Permitted Speed: 70 mph.

| 59701 | 04.91 | Battlefield Steam Railway | OP |

CLASS 115   TCL

Built: 1960 by BR at Derby C & W Works.
Length over buffers: 20.447 m.   **Weight:** 30.00 t.
Extreme width: 2.819 m.   **Seats:** 30 1st, 40 2nd.
Extreme height: 3.861 m.   **Toilet:** 1.
Maximum Permitted Speed: 70 mph.

| 59719 | 12.90 | South Devon Railway | OH |

CLASS 115   TS

Built: 1960 by BR at Derby C & W Works.
Length over buffers: 20.447 m.   **Weight:** 30.00 t.
Extreme width: 2.819 m.   **Seats:** 106 2nd.

Extreme height: 3.861 m.   **Toilet:** 1.
Maximum Permitted Speed: 70 mph.

59740	08.92	South Devon Railway	OH
59761	08.92	Buckinghamshire Railway Centre	UR

CLASS 107   TSL

Built: 1961 by BR at Derby C & W Works.
Length over buffers: 18.491 m.   **Weight:** 28.05 t.
Extreme width: 2.819 m.   **Seats:** 106 2nd.
Extreme height: 3.874 m.   **Toilet:** 1.
Maximum Permitted Speed: 70 mph.

59791	10.92	Battlefield Steam Railway	UR

CLASS 201   DMBS

Built: 1957 by BR at Eastleigh Works/Ashford Works.
Engine: English Electric type 4SRKT Mk.2 rated at 373 kW.
Transmission: Electric.
Length overall: 18.358 m.   **Weight:** 54.00 t.
Extreme width: 2.743 m.   **Seats:** 22.
Extreme height: 3.829 m.   **Toilet:** Not equipped.
Maximum Permitted Speed: 75 mph.

60000	09.86	Kent & East Sussex Railway	OP
60001	09.86	St. Leonards Railway Engineering	UR

CLASS 202   DMBS

Built: 1957 by BR at Eastleigh Works/Ashford Works.
Engine: English Electric type 4SRKT Mk.2 rated at 373 kW.
Transmission: Electric.
Length overall: 20.339 m.   **Weight:** 55.00 t.
Extreme width: 2.743 m.   **Seats:** 30.
Extreme height: 3.829 m.   **Toilet:** Not equipped.
Maximum Permitted Speed: 75 mph.

60016	09.86	Kent & East Sussex Railway	OP
60018	06.88	St. Leonards Railway Engineering	OP
60019	06.88	St. Leonards Railway Engineering	SU

CLASS 201   TSL

Built: 1957 by BR at Eastleigh Works/Ashford Works.

Length overall: 18.358 m. **Weight:** 29.00 t.
Extreme width: 2.743 m. **Seats:** 52.
Extreme height: 3.829 m. **Toilets:** 2.
Maximum Permitted Speed: 75 mph.

60500	09.86	St. Leonards Railway Engineering	SU
60501	09.86	St. Leonards Railway Engineering	SU
60502	09.86	St. Leonards Railway Engineering	SU

CLASS 202   TSL

Built: 1957 by BR at Eastleigh Works/Ashford Works.
Length overall: 20.339 m. **Weight:** 30.00 t.
Extreme width: 2.743 m. **Seats:** 60.
Extreme height: 3.829 m. **Toilets:** 2.
Maximum Permitted Speed: 75 mph.

60527	06.88	St. Leonards Railway Engineering	OP
60528	06.88	St. Leonards Railway Engineering	SU
60529	06.88	Kent & East Sussex Railway	OP

CLASS 201   TFLK

Built: 1957 by BR at Eastleigh Works/Ashford Works.
Length overall: 18.358 m. **Weight:** 30.00 t.
Extreme width: 2.743 m. **Seats:** 42.
Extreme height: 3.829 m. **Toilets:** 2.
Maximum Permitted Speed: 75 mph.

60700	09.86	St. Leonards Railway Engineering	SU

CLASS 201   TFLK

Built: 1957 by BR at Eastleigh Works/Ashford Works.
Length overall: 18.358 m. **Weight:** 31.00 t.
Extreme width: 2.743 m. **Seats:** 48.
Extreme height: 3.829 m. **Toilets:** 2.
Maximum Permitted Speed: 75 mph.

60708	09.86	St. Leonards Railway Engineering	UR
60709	06.88	St. Leonards Railway Engineering	SU

CLASS 201   TRB

Built: 1958 by BR at Eastleigh Works/Ashford Works.
Length overall: 18.358 m. **Weight:** 30.00 t.

Extreme width: 2.743 m. **Seats:** 21 unclassified.
Extreme height: 3.835 m. **Toilet:** 1 (Staff use only).
Maximum Permitted Speed: 75 mph.
Subsequent Number: DB 975386 in departmental service.

60750	01.64	St. Leonards Railway Engineering	SU

CLASS 204 DTCoL

Built: 1958 by BR at Eastleigh Works/Ashford Works.
Length overall: 20.358 m. **Weight:** 32.00 t.
Extreme width: 2.743 m. **Seats:** 13 1st, 62 2nd.
Extreme height: 3.823 m. **Toilets:** 2.
Maximum Permitted Speed: 75 mph.

60820	09.91	St. Leonards Railway Engineering	OP

UNCLASSIFIED DMBS

Built: 1954 by BR at Derby C & W Works.
Engines: Two horizontal AEC 220 type of 112 kW at 1800 rpm.
Transmission: Mechanical.
Length over buffers: 18.491 m. **Weight:** 27.00 t.
Extreme width: 2.794 m. **Seats:** 61.
Extreme height: 3.861 m. **Toilet:** 1.
Maximum Permitted Speed: 70 mph.
Subsequent Number: DB 975007 in departmental service.

79018	05.68	Mid Norfolk Railway	UR

CLASS 126 DMBSL

Built: 1956-57 by BR at Swindon Works.
Engines: Two horizontal AEC 220 type of 112 kW at 1800 rpm.
Transmission: Mechanical.
Length over buffers: 20.447 m. **Weight:** 38.85 t.
Extreme width: 2.819 m. **Seats:** 52.
Extreme height: 3.899 m. **Toilets:** 2.
Maximum Permitted Speed: 70 mph.

79091	10.72	LAMCO Railroad, Liberia	SU
79093	10.72	LAMCO Railroad, Liberia	SU
79094	10.72	LAMCO Railroad, Liberia	SU
79096	10.72	LAMCO Railraod, Liberia	SU
79097	10.72	LAMCO Railroad, Liberia	SU

CLASS 126   TBFKL

Built: 1957 by BR at Swindon Works.
Length over buffers: 20.155 m.   **Weight:**
Extreme width: 2.819 m.   **Seats:** 18.
Extreme height: 3.886 m.   **Toilets:** 2.
Maximum Permitted Speed: 70 mph.

79443	10.72	Bo'ness & Kinneil Railway	UR

UNCLASSIFIED   DTCL

Built: 1955 by BR at Derby C & W Works.
Length over buffers: 18.491 m.   **Weight:** 27.00 t.
Extreme width: 2.794 m.   **Seats:** 9 1st, 53 2nd.
Extreme height: 3.861 m.   **Toilet:** 1.
Maximum Permitted Speed: 70 mph.
Subsequent Number: DB 975008 in departmental service.

79612	01.68	Mid Norfolk Railway	UR

UNCLASSIFIED   DMS

Built: 1958 by Waggon und Maschinenbau, Donauwörth, West Germany.
Engine: One B•ssing type U10 of 112 kW at 1900 rpm.
Transmission: Mechanical.
Length over buffers: 13.950 m.   **Weight:** 15.00 t.
Extreme width: 2.648 m.   **Seats:** 56.
Extreme height: 3.607 m.   **Toilet:** Not equipped.
Maximum Permitted Speed: 55 mph.

79960	11.66	North Norfolk Railway	SU
79962	11.66	Keighley & Worth Valley Railway	SU
79963	11.66	North Norfolk Railway	OP
79964	04.67	Keighley & Worth Valley Railway	OP

UNCLASSIFIED   DMS

Built: 1958 by AC Cars, Thames Ditton.
Engine: One horizontal AEC 220 type of 112 kW at 1800 rpm.
Transmission: Mechanical.
Length over buffers: 11.328 m.   **Weight:** 11.00 t.
Extreme width: 2.819 m.   **Seats:** 46.
Extreme height: 3.721 m.   **Toilet:** Not equipped.
Maximum Permitted Speed: 58 mph.

79976	02.68	Mid Norfolk Railway	SU
79978	02.68	Colne Valley Railway	OP

2.2 "SECOND GENERATION" VEHICLES

CLASS R1   DMS

Built: 1978 by Leyland Bus, Workington/BREL, Derby C & W Works.
Engine: One horizontal Leyland 510 type of 149 kW at 1800 rpm.
Transmission: Mechanical.

Length overall: 12.400 m.	**Weight:** 14.00 t.
Extreme width: 2.500 m.	**Seats:** 44.
Extreme height: 3.900 m.	**Toilet:** Not equipped.

Maximum Permitted Speed: 75 mph.
Note: Originally an unpowered test vehicle. Power unit fitted 05.79.

975874   LEV 1   .78   National Railway Museum   SU

CLASS R3/1   DMS

Built: 1980 by Leyland Bus, Workington/D. Wickham, Ware.
Engine: One horizontal Leyland TL11 type of 149 kW at 1800 rpm.
Transmission: Mechanical.

Length over body: 15.300 m.	**Weight:** 19.40 t.
Extreme width: 2.970 m.	**Seats:** 56.
Extreme height: 3.965 m.	**Toilet:** Not equipped.

Maximum Permitted Speed: 75 mph.

LEV2   .81   Steamtown, Scranton, Pennsylvania, USA.   EX

CLASS R3/3   DMS

Built: 1981 by Leyland Bus, Workington/BREL Derby C & W Works.
Engine: One horizontal Leyland 690 type of 149 kW at 1800 rpm.
Transmission: Mechanical.

Length over body: 15.300 m.	**Weight:** 19.96 t.
Extreme width: 2.970 m.	**Seats:** 56.
Extreme height: 3.965 m.	**Toilet:** Not equipped.

Maximum Permitted Speed: 75 mph.

977020   R3   07.82   Northern Ireland Railways   OP

CLASS RB004   DMS

Built: 1984 by Leyland Bus, Workington/BREL Derby C & W Works.
Engine:
Transmission:

Length over body:	**Weight:**
Extreme width:	**Seats:**
Extreme height:	**Toilet:**
Maximum Permitted Speed:	
Un-numbered	Embsay Steam Railway

2.3 ADVANCED PASSENGER TRAIN VEHICLES

APT-E   PC

Built: 1972 by Metropolitan Cammell, Birmingham.
Power equipment: Four Leyland type 350 gas turbines, rated at 222 kW each.
Transmission: Electric.

Length overall: 23.000 m.	**Weight:** 49.00 t.
Extreme width: 2.685 m.	**Seats:** 0.
Extreme height: 3.640 m.	**Toilet:** Not equipped.
Maximum Permitted Speed:	

PC1	.76	National Railway Museum	CE
PC2	.76	National Railway Museum	CE

APT-E   TC

Built: 1972 by GEC Aircraft Division, Accrington.

Length overall:	**Weight:**
Extreme width: 2.685 m.	**Seats:**
Extreme height: 3.640 m.	**Toilet:**
Maximum Permitted Speed:	

TC1	.76	National Railway Museum	CE
TC2	.76	National Railway Museum	CE

2.4 FORMER GREAT WESTERN RAILWAY VEHICLES

UNCLASSIFIED   DMBT

Built: 1934 by Park Royal Vehicles, London.
Engines: Two AEC Ricardo 6 cylinder of 90 kW at 2000 rpm.
Transmission: Mechanical.

Length over buffers: 19.380 m.   **Weight:** 26.20 t.
Extreme width: 2.743 m.   **Seats:** 44.
Extreme height: 3.556 m.   **Toilets:** 2.
Maximum Permitted Speed: 70 mph.
W4W   4   07.58   Swindon Railway Museum   CE

UNCLASSIFIED   DMBT

Built: 1940 by Great Western Railway, Swindon.
Engines: Two AEC Ricardo 6 cylinder of 78 kW at 1650 rpm.
Transmission: Mechanical.
Length over buffers: 20.015 m.   **Weight:** 35.65 t.
Extreme width: 2.819 m.   **Seats:** 48.
Extreme height: 3.721 m.   **Toilets:**
Maximum Permitted Speed: 60 mph.
W20W   20   10.62   Kent & East Sussex Railway   UR

UNCLASSIFIED   DMBT

Built: 1940 by Great Western Railway, Swindon.
Engines: Two AEC Ricardo 6 cylinder of 78 kW at 1650 rpm.
Transmission: Mechanical.
Length over buffers: 20.015 m.   **Weight:** 35.65 t.
Extreme width: 2.819 m.   **Seats:** 48.
Extreme height: 3.718 m.   **Toilet:** Not equipped.
Maximum Permitted Speed: 40 mph.
W22W   22   10.62   Didcot Railway Centre   OP

PART THREE - ELECTRIC MULTIPLE UNITS

3.1 POST NATIONALISATION STOCK

CLASS 370 (APT-P)   6-Car Articulated Unit

Built: 1978 by BREL, Derby Works (trailer cars) or Crewe Works (power cars).
Formation: DTSO+TBFO+M+TRSB+TBFO+DTSO.
Supply System: 25 kV ac 50Hz overhead.
Traction Motors: Four ASEA LJMA 410F body mounted per power car.
Length overall: **Weight:** 292.95 t.
Extreme width: 2.724 m.   **Seats:** 50 1st, 132 2nd.

Maximum Permitted Speed: 125 mph. **Toilets:** 5.

48103	DTSO	09.88	The Railway Age, Crewe	CE
48106	DTSO	09.88	The Railway Age, Crewe	CE
48602	TBFO	09.88	The Railway Age, Crewe	CE
48603	TBFO	09.88	The Railway Age, Crewe	CE
48404	TRSB	09.88	The Railway Age, Crewe	CE
49002	M	09.88	The Railway Age, Crewe	CE
49006	M	09.88	ABB Transportation, Crewe	UR

CLASS 405   4-Sub TSO

Built: 1948 by BR at Eastleigh Works.
Length over body: 18.900 m.   **Weight:** 26.00 t.
Extreme width: 2.819 m.   **Seats:** 102.
Maximum Permitted Speed: 75 mph. **Toilet:** Not equipped.

12354	.94	BR Brighton Depot	ML

CLASS 405   4-Sub DMBSO

Built: 1937 by BR at Eastleigh Works.
Supply System: 750 V dc third rail.
Traction Motors: Two EE507.
Length over body: 19.050 m.   **Weight:** 42.00 t.
Extreme width: 2.819 m.   **Seats:** 82.
Maximum Permitted Speed: 75 mph. **Toilet:** Not equipped.

12795	.94	BR Brighton Depot	ML
12796	.94	BR Brighton Depot	ML

CLASS 491 (4-TC)   4-Car Trailer Unit

Built: 1966-67 (* 1974) by BR at York Works.
Formation: DTSO+TBSK+TFK+DTSO.
Length overall: 80.950 m.   **Weight:** 132.00 t.
Extreme width: 2.819 m.   **Seats:** 50 1st, 160 2nd.
Maximum Permitted Speed: 90 mph. **Toilets:** 3.

70823	TBSO	05.91	London Underground Ltd, Ruislip	OH
70824	TBSO	02.92	London Underground Ltd, Ruislip	OH
70855	TFK	05.91	London Underground Ltd, Ruislip	OH
71163*	TFK	02.92	London Underground Ltd, Ruislip	OH
76297	DTSO	05.91	London Underground Ltd, Ruislip	OH
76298	DTSO	05.91	London Underground Ltd, Ruislip	OH
76322	DTSO	02.92	London Underground Ltd, Ruislip	OH
76324	DTSO	02.92	London Underground Ltd, Ruislip	OH

CLASS 485/486   Isle of Wight Stock DMBSO

Built: 1931-32 by Metropolitan Cammell, Birmingham for London Electric Railway. Purchased by BR and converted in 1967 for use on the Isle of Wight.
Supply System: 630 V dc third rail.
Traction Motors: Two GEC WT54A per car.
Length overall: 16.190 m.   **Weight:** 31.75 t.
Extreme width: 2.690 m.   **Seats:** 26.
Maximum Permitted Speed: 45 mph.   **Toilet:** Not equipped.

2	11.90	London Underground Ltd, Acton Works	SU
7	11.90	London Underground Ltd, Acton Works	SU

CLASS 485/486   Isle of Wight Stock DTSO

Built: 1925 by Metropolitan Cammell, Birmingham for London Electric Railway. Purchased by BR and converted in 1967 for use on the Isle of Wight.
Length overall: 15.820 m.   **Weight:** 17.05 t.
Extreme width: 2.690 m.   **Seats:** 38.
Maximum Permitted Speed: 45 mph.   **Toilet:** Not equipped.

27	11.90	London Underground Ltd, Acton Works	UR

CLASS 485/486   Isle of Wight Stock TSO

Built: 1924 by Cammell Laird, Birkenhead for London Electric Railway. Purchased by BR and converted in 1967 for use on the Isle of Wight.
Length overall: 15.690 m.   **Weight:** 18.95 t.
Extreme width: 2.690 m.   **Seats:** 42.
Maximum Permitted Speed: 45 mph.   **Toilet:** Not equipped.

44	11.90	London Underground Ltd, Acton Works	SU
49	11.90	London Underground Ltd, Acton Works	UR

CLASS 501   Watford TSO

Built: 1957 by BR at Eastleigh Works.
Length: 17.42 m.   **Weight:** 29.50 t.
Extreme width: 2.819 m.   **Seats:** 92.
Maximum Permitted Speed: 60 mph.   **Toilet:** Not equipped.

70170	10.85	MOD Marchwood Port Railway	OH

CLASS 501   Watford DTBSO

Built: 1957 by BR at Eastleigh Works.

Length: 17.52 m.		**Weight:** 30.50 t.		
Extreme width: 2.819 m.		**Seats:** 74.		
Maximum Permitted Speed: 60 mph.		**Toilet:** Not equipped.		
75186		10.85	MOD Marchwood Port Railway	OH

CLASS 504 — Bury 2-Car Unit

Built: 1959 by BR at Wolverton Works.
Formation: DMBSO+DTSO.
Supply System: 1200 V dc third rail side contact.
Traction Motors: Four English Electric.

Length: 20.31 + 20.31 m.		**Weight:** 82.00 t.	
Extreme width: 2.819 m.		**Seats:** 178.	

Maximum Permitted Speed: 65 mph. **Toilet:** Not equipped.

65451	DMBSO	08.91	East Lancashire Railway	SU
77172	DTSO	08.91	East Lancashire Railway	SU

CLASS 506 — Hadfield 3-Car Unit

Built: 1950 by Metropolitan Cammell, Birmingham (* Birmingham Railway Carriage & Wagon Company, Smethwick).
Formation: DMBSO+TSO+DTSO.
Supply System: 1500 V dc overhead.
Traction Motors: Four GEC.

Length: 18.41 + 16.78 + 16.87 m.		**Weight:** 106.00 t.	
Extreme width: 2.819 m.		**Seats:** 174.	

Maximum Permitted Speed: 70 mph. **Toilet:** Not equipped.

59404	DMBSO	12.84	Midland Railway Centre	CE
59504	TSO	12.84	Midland Railway Centre	CE
59604	DTSO	12.84	Midland Railway Centre	CE

UNCLASSIFIED — 4-DD DMBSO

Built: 1949 by BR at Lancing Works.
Supply System: 750 V dc third rail.
Traction Motors: Two English Electric.

Length over body: 19.050 m.		**Weight:** 39.00 t.	
Extreme width: 2.819 m.		**Seats:** 120.	

Maximum Permitted Speed: 75 mph. **Toilet:** Not equipped.

13003		10.71	Great Bower Farm, near Challock	SU
13004		10.71	Northampton & Lamport Railway	UR

UNCLASSIFIED DMBS

Built: 1958 by BR at Derby/Cowlairs Works.
Supply System: 216 lead acid batteries of 1070 Ah capacity.
Traction Motors: Two Siemens-Schuckert.
Length overall: 18.491 m.   **Weight:** 37.50 t.
Extreme width: 2.794 m.   **Seats:** 52.
Maximum Permitted Speed: 70 mph.   **Toilet:** Not equipped.
Subsequent Number: DB 975003 in departmental service

79998	12.66	East Lancashire Railway	OP

UNCLASSIFIED DTCL

Built: 1958 by BR at Derby/Cowlairs Works.
Length overall: 18.491 m.   **Weight:** 37.50 t.
Extreme width: 2.794 m.   **Seats:** 12 1st, 53 2nd.
Maximum Permitted Speed: 70 mph.   **Toilet:** 1.
Subsequent Number: DB 975004 in departmental sevice

79999	12.66	East Lancashire Railway	OP

3.2 FORMER LMS & CONSTITUENT COMPANIES STOCK

CLASS 502 Southport 2-Car Unit

Built: 1939 by London Midland & Scottish Railway, Derby C & W Works.
Formation: DMBSO+DTSO.
Supply System: 630 V dc third rail.
Traction Motors: Four English Electric.
Length: 20.26 + 20.26 m.   **Weight:** 68.00 t.
Extreme width: 2.900 m.   **Seats:** 167.
Maximum Permitted Speed: 65 mph.   **Toilet:** Not equipped.

28361	DMBSO	08.80	Southport Railway Centre	SS
29896	DTSO	08.80	Southport Railway Centre	SS

CLASS 503 Wirral 3-Car Unit

Built: 1938 by Metropolitan Cammell, Birmingham (*Birmingham Railway Carriage & Wagon Company, Smethwick).
Formation: DMBSO+TSO+DTSO.

Battery EMU 79998 + 79999 at the now demolished Bradford Hammerton Street depot early in 1989.
(Neil Webster)

Supply System: 650 V dc third rail.
Traction Motors: Four BTH.
Length: 17.68 + 17.07 + 17.68 m.   **Weight:** 78.00 t.
Extreme width: 2.77 m.   **Seats:** 176.
Maximum Permitted Speed: 65 mph.   **Toilet:** Not equipped.

28690*	DMBSO	03.85	Wirral MDC at Kirkdale EMU Depot	SU
29289	DTSO	03.85	Wirral MDC at Kirkdale EMU Depot	SU
29720	TSO	03.85	Wirral MDC at Kirkdale EMU Depot	SU

CLASS 505   Altrincham TC

Built: 1931 by Metropolitan Cammell, Birmingham.
Length over body: 17.399 m.   **Weight:** 30.00 t.
Extreme width: 2.819 m.   **Seats:** 24 1st, 72 2nd.
Maximum Permitted Speed: mph.   **Toilet:** Not equipped.

29663	05.71	Midland Railway Centre	CE
29666	05.71	Midland Railway Centre	CE
29670	05.71	Midland Railway Centre	CE

UNCLASSIFIED   Watford DMBSO

Built: 1915 by Metropolitan Cammell, Birmingham.
Supply System: 630 V dc third rail.
Traction Motors: Four Oerlikon.
Length over body: 17.600 m.   **Weight:** 54.75 t.
Extreme width: 2.73 m.   **Seats:** 48.
Maximum Permitted Speed: mph.   **Toilet:** Not equipped.

28249	05.62	National Railway Museum	CE

3.3 FORMER SR STOCK

CLASS 487   Waterloo & City DMBSO

Built: 1940 by English Electric, Preston.
Supply System: 630 V dc third rail.
Traction Motors: Two English Electric 500.
Length overall: 14.860 m.   **Weight:** 29.00 t.
Extreme width: 2.640 m.   **Seats:** 40.
Maximum Permitted Speed: 35 mph.   **Toilet:** Not equipped.

61	05.93	National Railway Museum	CE

†285 CLASS   3-Sub DMBS

Built: 1925 by Metropolitan Cammell, Birmingham.
Supply System: 750 V dc third rail.
Traction Motors: Two Metropolitan Vickers.
Length overall: **Weight:**
Extreme width: **Seats:**
Maximum Permitted Speed: 75 mph. **Toilet:** Not equipped.

8143	11.60	National Railway Museum	CE

UNCLASSIFIED   4-Cor TSK

Built: 1938 by Southern Railway, Eastleigh Works.
Length over body: 19.355 m.   **Weight:** 32.65 t.
Extreme width: 2.819 m.   **Seats:** 68.
Maximum Permitted Speed: 75 mph. **Toilet:** 1.

10096	12.72	St. Leonards Railway Engineering	UR

CLASS 405   4-Sub TS

Built: 1947 by Southern Railway, Eastleigh Works.
Length overall: 18.90 m.   **Weight:** 27.00 t.
Extreme width: 2.819 m.   **Seats:** 120.
Maximum Permitted Speed: 75 mph. **Toilet:** Not equipped.

10239	.94	BR Brighton Depot	ML

UNCLASSIFIED   2-Bil DTCL

Built: 1937 by Southern Railway, Eastleigh Works.
Length over body: 19.050 m.   **Weight:** 31.25 t.
Extreme width: 2.819 m.   **Seats:** 24 1st, 32 2nd.
Maximum Permitted Speed: 75 mph. **Toilet:** 1.

10656	08.71	BR Brighton Depot	SS

UNCLASSIFIED   4-Res DMBSO

Built: 1937 by Southern Railway, Eastleigh Works.
Supply System: 750 V dc third rail.
Traction Motors: Two English Electric.
Length over body: 19.355 m.   **Weight:** 46.50 t.
Extreme width: 2.858 m.   **Seats:** 52.
Maximum Permitted Speed: 75 mph. **Toilet:** Not equipped.

UNCLASSIFIED   4-Cor DMBSO

Built: 1937-38 by Southern Railway, Eastleigh Works.
Supply System: 750 V dc third rail.
Traction Motors: Two English Electric.
Length over body: 19.355 m. **Weight:** 46.50 t.
Extreme width: 2.858 m. **Seats:** 52.
Maximum Permitted Speed: 75 mph. **Toilet:** Not equipped.

11179	09.72	National Railway Museum	CE
11187	09.72	St. Leonards Railway Engineering	UR
11201	09.72	St. Leonards Railway Engineering	UR

UNCLASSIFIED   6-Pul TCK

Built: 1932 by Southern Railway, Eastleigh Works.
Length over body: 19.355 m. **Weight:** 32.60 t.
Extreme width: 2.819 m. **Seats:** 30 1st, 24 2nd.
Maximum Permitted Speed: 75 mph. **Toilet:** 1.

11773	.72	Swanage Railway

UNCLASSIFIED   4-Cor TCK

Built: 1937 by Southern Railway, Eastleigh Works.
Length over body: 19.355 m. **Weight:** 32.60 t.
Extreme width: 2.819 m. **Seats:** 30 1st, 24 2nd.
Maximum Permitted Speed: 75 mph. **Toilet:** 1.

11825	.72	St. Leonards Railway Engineering

UNCLASSIFIED   2-Bil DMBSOL

Built: 1937 by Southern Railway, Eastleigh Works.
Supply System: 750 V dc third rail.
Traction Motors: Two English Electric.
Length over body: 19.050 m. **Weight:** 43.50 t.
Extreme width: 2.819 m. **Seats:** 52.
Maximum Permitted Speed: 75 mph. **Toilet:** 1.

12123	08.71	BR Brighton Depot	SS

GRIMSBY & IMMINGHAM LIGHT RAILWAY TRAM

Built: 1914 by Great Central Railway, Dukinfield Works.
Supply System:
Traction Motors: Two Dick Kerr.
Length over body: **Weight:**
Extreme width: **Seats:** 64.
Maximum Permitted Speed: **Toilet:** Not equipped.

14	07.61	National Tramway Museum

GRIMSBY & IMMINGHAM LIGHT RAILWAY TRAMS

Built: 1925-7 by Gateshead & District Tramways Company, Gateshead.
Supply System:
Traction Motors:
Length over body: **Weight:**
Extreme width: **Seats:** 48.
Maximum Permitted Speed: **Toilet:** Not equipped.

20	07.61	National Tramway Museum	
26	07.61	National Tramway Museum	OP

UNCLASSIFIED   North Tyneside DMPV

Built: 1904 by North Eastern Railway, York Works.
Supply System: 675 V dc third rail.
Traction Motors:
Length over body: 17.400 m.   **Weight:** 46.00 t.
Extreme width: 2.772 m.   **Seats:** Not equipped.
Maximum Permitted Speed: **Toilet:** Not equipped.

900730	.38	North Tyneside Steam Railway	CE

3.5 FORMER PULLMAN CAR COMPANY STOCK

UNCLASSIFIED   6-Pul TPCK

Built: 1932 by Metropolitan Cammell, Birmingham.
Length over body: 20.400 m.   **Weight:** 43.00 t.
Extreme width: 2.772 m.   **Seats:**
Maximum Permitted Speed: 75 mph. **Toilet:**

264	Ruth	09.65	VSOE at BR Stewarts Lane Depot	
278	Bertha	06.66	Bluebell Railway	OH

UNCLASSIFIED   5-Bel TPKF

Built: 1932 by Metropolitan Cammell, Birmingham.
Length over body: 20.117 m.   **Weight:** 43.00 t.
Extreme width: 2.731 m.   **Seats:** 20.
Maximum Permitted Speed: 75 mph. **Toilet:**

279	Hazel	04.72	The Black Bull, Moulton	CE
280	Audrey	04.72	VSOE (at BR Stewarts Lane Depot)	ML
281	Gwen	04.72	VSOE (at BR Stewarts Lane Depot)	
282	Doris	04.72	CIL Shopfitters, Finsbury Park	CE
283	Mona	04.72	The Brighton Belle, Winsford	CE
284	Vera	04.72	VSOE (at BR Stewarts Lane Depot)	ML

UNCLASSIFIED   5-Bel TPS

Built: 1932 by Metropolitan Cammell, Birmingham.
Length over body: 20.117 m.   **Weight:** 39.00 t.
Extreme width: 2.731 m.   **Seats:** 56.
Maximum Permitted Speed: 75 mph. **Toilet:**

85		04.72	Current location unknown	
86		04.72	VSOE (at BR Stewarts Lane Depot)	
87		04.72	North Norfolk Railway	OH

UNCLASSIFIED   5-Bel DMPBS

Built: 1932 by Metropolitan Cammell, Birmingham.
Supply System: 750 V dc third rail.

Traction Motors: Four BTH.
Length over body: 20.117 m.   **Weight:** 62.00 t.
Extreme width: 2.731 m.   **Seats:** 48.
Maximum Permitted Speed: 75 mph. **Toilet:**

88	04.72	VSOE (at BR Stewarts Lane Depot)	
89	04.72	Little Mill Inn, Rowarth	CE
90	04.72	East Lancashire Railway	SU
91	04.72	North Norfolk Railway	OH
92	04.72	Brighton Railway Museum	SU
93	04.72	Brighton Railway Museum	SU

LOCOMOTIVE NAMES

Names are listed in the same order as the vehicles are listed. Names bestowed since acquisition from BR are shown in *italic* typeface.

D2867	*Diane*
D2868	*Sam*
D2023	*Faith*
03059	*Edward*
03119	*Linda*
03144	*Western Waggoner*
03162	Birkenhead South 1879-1985
D2192	*Ardent*
D2272	*Alfie*
D2578	*Cider Queen*
07011	*Cleveland*
D3002	*Dulcote*
08011	*Haversham*
D3019	*Gwyneth*
08022	*Lion*
08032	*Mendip*
08046	*Brechin City*
08060	*Unicorn*
08077	*James*
08123	*George Mason*
08164	*Prudence*
08320	*Susan*
08331	*Terence*
08398	*Annabel*
08678	*Ulverstonian*
08764	*Florence*
08774	*Arthur Vernon Dawson*
08785	*Clarence*
D3489	*Colonel Tomline*

D4067	_Margaret Ethel-Thomas Alfred Naylor_
D4092	_Christine_
20041	_Nancy_
20060	_Lorna_
20083	_Alison_
20101	_Janis_
20219	_Kilmarnock 400_
20225	_Iona_
20227	_Traction_
24032	_Helen Turner_
24061	Experiment
25035	_Castel Dinas Bram_
25173	_John F. Kennedy_
25185	_Mercury_
25191	_The Diana_
25265	_Harlech Castle/Castell Harlech_
25278	_Sybillia_
25322	Tamworth Castle
33056	The Burma Star
33114	Ashford 150
D7017	_Williton_
37190	Dalzell
40012	Aureol
40013	Andania
40106	_Atlantic Conveyor_
D821	Greyhound
D832	Onslaught
44004	Great Gable
44008	Penyghent
45060	Sherwood Forester
45112	The Royal Army Ordnance Corps
45118	The Royal Artilleryman
45135	3rd Carabinier
46035	Ixion
47401	North Eastern (12.81 to 05.88). Star of the East (05.91 to 06.92)
47402	Gateshead
50002	Superb
50007	Sir Edward Elgar
50008	Thunderer
50015	Valiant
50017	Royal Oak
50019	Ramilles
50021	Rodney
50026	Indomitable
50027	Lion
50031	Hood

50033	Glourious
50035	Ark Royal
50042	Triumph
50043	Eagle
50044	Exeter
50049	Defiance
50050	Fearless
D1010	Western Campaigner
D1013	Western Ranger
D1015	Western Champion
D1023	Western Fusilier
D1041	Western Prince
D1048	Western Lady
D1062	Western Courier
55002	The King's Own Yorkshire Light Infantry
55009	Alycidon
55015	Tulyar
55016	Gordon Highlander
55019	Royal Highland Fusilier
55022	Royal Scots Grey
E27000	Electra
E27001	Ariadne
E27003	Diana
89001	Avocet
7050	*Rorke's Drift*
7051	*John Alcock*

LIST OF INDUSTRIAL USERS

Please note that the addresses listed below are private sites and many are not normally open to the public without prior written permission.

ACCIAIERIE.
Lonato, ITALY.

ABB TRANSPORTATION, CREWE.
West Street, Crewe, Cheshire.

ABB TRANSPORTATION, DERBY C & W.
Litchurch Lane, Derby, Derbyshire.

ABB TRANSPORTATION, YORK.
Holgate Road, York, North Yorkshire.

ANGLO-CHARRINGTON.
Pensnett Fuel Terminal, Shutt End, Brierley Hill, West Midlands.

BCOE, OXCROFT DISPOSAL POINT.
Stanfree, near Clowne, Derbyshire.

BOKE RAILWAY.
Chemin de fer de BOKE, Conakry, GUINEA, West Africa.

BRITISH FUEL CO.
Whalley Banks, Blackburn, Lancashire.

BRITISH SALT, MIDDLEWICH.
Cledford Lane, Middlewich, Cheshire.

BRML, EASTLEIGH.
British Rail Maintenance Limited, Campbell Road, Eastleigh, Hampshire.

H.P. BULMER.
Plough Lane, Hereford, Hereford & Worcester.

BUTTERLEY ENGINEERING CO.
Ripley, Derbyshire.

CAMAS AGGREGATES.
Meldon Quarry, Okehampton, Devon.

CARBIBONI.
Carbiboni PW Contractors, Colico, ITALY.

CIL SHOPFITTERS.
Fonthill Road, Finsbury Park, Greater London.

COURTAULDS, BRIDGWATER.
Courtaulds Ltd., Films & Packaging, Bridgwater, Somerset.

CFD, AUTUN, FRANCE.
CFD Industrie, rue Benôit-Malon, 92150 Suresnes, FRANCE.

COBRA, WAKEFIELD.
Cobra Railfreight Ltd., Calder Vale Road, Wakefield, West Yorkshire.

COSTAIN DOW-MAC, TALLINGTON.
Barholm Road, Tallington, near Stamford, Lincolnshire.

CO-STEEL, SHEERNESS.
Sheerness on Sea, Kent.

A.V. DAWSON, MIDDLESBROUGH.
Depot Road, Middlesborough, Cleveland.

DAY AGGREGATES, BRENTFORD.
Brentford Town Goods, Transport Avenue, Brentford, Greater London.

DEANSIDE TRANSIT, GLASGOW.
Deanside Road, Hillington, Glasgow, Strathclyde.

J.M. DEMULDER, SHILTON.
Gun Range Farm, Shilton Lane, Shilton, near Coventry, Warwickshire.

ENGLISH CHINA CLAYS, BURNGULLOW.
Burngullow, St Austell, Cornwall.

ENGLISH CHINA CLAYS, ROCKS DRIERS.
Goonbarrow, Bugle, Cornwall.

FELIXSTOWE DOCK & RAILWAY COMPANY.
European House, The Dock, Felixstowe, Suffolk.

FIRE SERVICES TRAINING CENTRE, MORETON IN MARSH.
The Fire Service College. Moreton in Marsh, Gloucestershire.

FORD MOTOR COMPANY, DAGENHAM.
Dagenham, Greater London.

FOSTER YEOMAN, MEREHEAD.
Foster Yeoman Quarries Ltd., Torr Works, Shepton Mallet, Somerset.

FOSTER YEOMAN, ISLE OF GRAIN.
Foster Yeoman Quarries Ltd., Isle of Grain Site, Rochester, Kent.

GENAPPE SUGAR FACTORY.
Charleroi, BELGIUM.

GULF OIL, WATERSTON.
Gulf Oil Refining Ltd., Waterston, near Milford Haven, Dyfed.

ARTHUR GUINNESS, PARK ROYAL.
Arthur Guiness, Son & Co (Park Royal) Ltd., Park Royal Brewery, Greater London.

HUMBERSIDE SEA & LAND SERVICES, GRIMSBY.
Grimsby, Humberside.

HUNSLET-BARCLAY LTD.
Caledonia Works, West Longlands Street, Kilmarnock, Strathclyde.

ICI, WILTON.
ICI Chemicals & Polymers Ltd, Wilton Works, Middlesbrough, Cleveland.

ISA OSPITALETTO.
ISA Ospitaletto, Brescia, ITALY.

LAMCO RAILROAD.
Liberian American Swedish Minerals Co, Nimba Workshops, Roberts Airport, LIBERIA.

LOCOTEC
Vanguard Works, Bretton Street, Dewsbury, West Yorkshire.

LONDON UNDERGROUND Ltd., ACTON.
Acton Works, Bollo Lane, Acton, Greater London.

LONDON UNDERGROUND LTD, RUISLIP.
West Ruislip Depot, Ruislip, Greater London.

MAYER PARRY RECYCLING.
Scrubs Lane, Willesden, Greater London.

MAYER PARRY RECYCLING.
Snailwell, near Newmarket, Cambridgeshire

M.C. METAL PROCESSING, GLASGOW.
Springburn Works, Springburn Road, Glasgow, Strathclyde.

MOD, BICESTER.
Bicester Millitary Railway, Bicester, Warwickshire.

MOD, LONG MARSTON.
Engineer Resources, Long Marston, Warwickshire.

MOD, LUDGERSHALL.
Central Vehicle Depot, Ludgershall, Wiltshire.

MOD, MARCHWOOD.
Marchwood Military Port, Hampshire.

OTIS EURO TRANSRAIL, SALFORD.
Liverpool Road, Salford, Greater Manchester.

PD FUELS, COED BACH DP.
Powell Duffryn Coal Preparation Ltd., Coed Bach Disposal Point, Kidwelly, Dyfed.

PD FUELS, GWAUN-CAE-GURWEN DP.
Powell Duffryn Coal Preparation Ltd., Gwaun-Cae-Gurwen Disposal Point, Gwaun-Cae-Gurwen, West Glamorgan.

POTTER GROUP, ELY.
Queen Adelaide, Ely, Cambridgeshire.

REDLAND ROADSTONE, BARROW ON SOAR.
Redland Aggregates Ltd., Barrow on Soar, Leicestershire.

RFS(E) Ltd.
Doncaster Works, Hexthorpe Road, Doncaster, South Yorkshire.
SIDERURGICA MONTIRONE.
Brescia, Italy.

TILCON, GRASSINGTON.
Swinden Lime Works, Grassington, near Skipton, North Yorkshire.

TRAFFORD PARK ESTATES, MANCHESTER.
Twining Road, Ashburton, Trafford Park, Manchester, Gtr. Manchester.

LIST OF PRESERVATION SITES

Please note that whilst the majority of sites shown below are open to the public, days/hours of opening may be restricted and potential visitors are advised to telephone in advance for details.

AVON & RADSTOCK RAILWAY.
Radstock, Avon.

AVON VALLEY RAILWAY.
Bitton Station, Willsbridge, Bristol, Avon, BS15 6ED. (0117 932 7296)

BATTLEFIELD STEAM RAILWAY.
Shackerstone Station, near Market Bosworth, Leicestershire, CV13 8NW. (01827 880754)

BIRMINGHAM RAILWAY MUSEUM.
670 Warwick Road, Tyseley Depot, Birmingham, West Midlands, B11 2HL. (0121 707 4696)

BLACK BULL, THE
Moulton, near Richmond, North Yorkshire, DL10 6QJ. (01325 377289)

BLUEBELL RAILWAY.
Sheffield Park Station, Uckfield, East Sussex, TN22 3QL. (01825 723777)

BNFL, Salwick.
SHAME, c/o British Nuclear Fuels Ltd, Springfields Factory, Salwick, Preston, Lancashire.

BODMIN & WENFORD RAILWAY.
Bodmin General Station, Bodmin, Cornwall, PL31 1AQ. (01208 73666)

BO'NESS & KINNEIL RAILWAY.
Bo'ness Station, Union Street, Bo'ness, Central, EH51 0AD. (01506 822298)

BRIGHTON BELLE, THE
Middlewich Road, Winsford, Cheshire, CW7 3NQ. (01606 593292)

BRIGHTON RAILWAY MUSEUM.
Preston Park, Brighton, West Sussex.

BSC, SCUNTHORPE.
Traffic Department, British Steel PLC, PO Box 1, Scunthorpe, Humberside, DN16 1BP.

BUCKINGHAMSHIRE RAILWAY CENTRE.
The Railway Station, Quainton, near Aylesbury, Buckinghamshire, HP22 4BY. (01296 655720)

CADEBY LIGHT RAILWAY.
The Old Rectory, Cadeby, Nuneaton, Warwickshire, CV13 0AS. (01455 290462)

CAERPHILLY RAILWAY SOCIETY.
Harold Wilson Industrial Estate, Van Road, Caerphilly, Mid Glamorgan. (01222 461414)

CALEDONIAN RAILWAY.
The Station, 2 Park Road, Brechin, Tayside, DD9 7AF. (01345 55965 or 01674 81318)

CAMBRIAN RAILWAYS SOCIETY.
Oswestry Station Yard, Oswald Road, Oswestry, Shropshire. (01691 661648)

CHASEWATER LIGHT RAILWAY.
Chasewater Park, Brownhills, Staffordshire, WS8 7LT. (01543 452623)

CHEDDLETON RAILWAY CENTRE.
Cheddleton Station, near Leek, Staffordshire, ST13 7EE. (01538 360522)

CHINNOR & PRINCESS RISBOROUGH RAILWAY.
Chinnor Cement Works, Chinnor, Oxfordshire. (01296 433795)

CHOLSEY & WALLINGFORD RAILWAY.
Hithercroft Industrial Estate, Wallingford, Oxfordshire, OX10 0NF. (01491 835067)

COLNE VALLEY RAILWAY.
Castle Hedingham Station, Yeldham Road, Castle Hedingham, Halsted, Essex, CO9 3DZ. (01787 461174)

DARLINGTON RAILWAY CENTRE & MUSEUM.
North Road Station, Darlington, County Durham, DL3 6ST. (01325 460532)

DEAN FOREST RAILWAY.
Norchard Steam Centre, Lydney, Gloucestershire, GL15 4ET. (01594 845840 or 01452 840625)

DIDCOT RAILWAY CENTRE.
Didcot, Oxfordshire, OX11 7NJ. (01235 817200)

EAST ANGLIAN RAILWAY MUSEUM
Chappel & Wakes Colne Station, near Colchester, Essex, CO6 2DS. (01206 242524)

EAST KENT LIGHT RAILWAY.
Shepherdswell (EKLR) Station, Shepherdswell, near Dover, Kent. (01304 822850)

EAST LANCASHIRE RAILWAY.
Bolton Street Station, Bury, Greater Manchester, BL9 0EY. (0161.764 7790)

EMBSAY STEAM RAILWAY.
Embsay Station, Embsay, Skipton, North Yorkshire, BD23 6AX. (01756 794727)

FAWLEY HILL RAILWAY.
Private Site near Henley on Thames.

FSAS, ARREZO.
Stia-Arrezo-Sinalunga Railway, Conicno 2, Arrezo, ITALY.

GLOUCESTERSHIRE WARWICKSHIRE RAILWAY.
The Station, Toddington, Cheltenham, Gloucestershire, GL54 5DT. (01242 621405)

GREAT BOWER FARM.
Private Site near Challock, Kent.

GREAT CENTRAL RAILWAY.
Central Station, Great Central Road, Loughborough, Leicestershire, LE11 1RW. (01509 230726)

GREAT CENTRAL RAILWAY - RUDDINGTON.
Notingham Heritage Centre, Mere Way, Ruddington, Nottingham, NG11 6NX. (0115 940 5705)

GRIMSBY & LOUTH RAILWAY.
Ludborough Station, Lincolnshire.

GWILI RAILWAY/RHEILFFORDD GWILI.
Bronwydd Arms Station, Bronwydd Arms, Carmarthen, Dyfed, SA33 6HT. (01267 230666)

HERITAGE & MARITIME MUSEUM.
Old Custom House, Milford Dock, Milford Haven, Dyfed. (01646 694496)

HUMBERSIDE LOCOMOTIVE PRESERVATION GROUP.
c/o BR Dairycoates Depot, Hull, Humberside.

IRCHESTER NARROW GAUGE RAILWAY MUSEUM.
Irchester Country Park, Irchester, Northamptonshire. (01604 844763)

ISLE OF WIGHT STEAM RAILWAY.
Haven Street Station, Ryde, Isle of Wight, PO33 4DS. (01983 882204)

KEIGHLEY & WORTH VALLEY RAILWAY.
Haworth Station, Keighley, West Yorkshire, BD22 8NJ. (01535 645214)

KENT & EAST SUSSEX STEAM RAILWAY.
Tenterden Town Station, Tenterden, Kent, TN30 6HE. (01580 765155)

LAKESIDE & HAVERTHWAITE RAILWAY.
Haverthwaite Station, near Ulverston, Cumbria, LA12 8AL. (01539 531594)

LAVENDER LINE.
Isfield Station, Isfield, near Uckfield, East Sussex, TN22 5XB. (01825 750515)

LITTLE MILL INN.
Rowarth, Derbyshire, SK12 5EB. (01663 743178)

LLANGOLLEN RAILWAY.
Llangollen Station, Abbey Road, Llangollen, Clwyd, LL20 8SN. (01978 860979)

MANGAPPS FARM RAILWAY.
Southminster Road, Burnham-on-Crouch, Essex, CM0 8QQ. (01621 784898)

MIDDLETON RAILWAY.
Moor Road, Hunslet, Leeds, West Yorkshire, LS10 2JQ. (0113 271 0320)

MID HANTS RAILWAY.
Alresford Station, Alresford, Hampshire, SO24 9JG. (01962 733810/734200)

MIDLAND RAILWAY CENTRE.
Butterley Station, near Ripley, Derbyshire, DE5 3TL. (01773 747674/570140)

MID NORFOLK RAILWAY.
County School Station, North Eltham, Dereham, Norfolk. (01362 668181)

MUSEUM OF ARMY TRANSPORT.
Flemingate, Beverley, North Humberside, HU17 0NG. (01482 860445)

MUSEUM OF SCIENCE & INDUSTRY, MANCHESTER.
Liverpool Road Station, Castlefield, Manchester, Greater Manchester, M3 4JP.
(0161 832 2244)

NARROW GAUGE RAILWAY CENTRE.
Gloddfa Ganol Slate Mine, Blaenau Ffestiniog, Gwynedd, LL41 3NB. (01766 830664)

NATIONAL RAILWAY MUSEUM.
Leeman Road, York, North Yorkshire, YO2 0XJ. (01904 621261)

NATIONAL TRAMWAY MUSEUM.
Crich, near Matlock, Derbyshire, DE4 5DP. (01773 852565)

NENE VALLEY RAILWAY.
Wansford Station, Stibbington, Peterborough, Cambridgeshire, PE8 6LR.
(01780 782854)

NORTHAMPTON & LAMPORT RAILWAY.
Pitsford & Brampton Station, Pitsford Road, Chapel Brampton, Northamptonshire,
NN6 8BA. (01604 22709)

NORTH NORFOLK RAILWAY.
Sheringham Station, Sheringham, Norfolk, NR26 8RA. (01263 822045)

NORTH TYNESIDE STEAM RAILWAY.
Middle Engine Lane, West Chirton, North Shields, Tyne & Wear, NE29 8DX. (0191 262 2627)

NORTH YORKSHIRE MOORS RAILWAY.
Pickering Station, Pickering, North Yorkshire, YO18 7AJ. (01751 472508/473535)

PAIGNTON & DARTMOUTH STEAM RAILWAY.
Queen's Park Station, Torbay Road, Paignton, Devon, TQ4 6AF. (01803 555872)

PEAK RAIL.
Matlock Station, Matlock, Derbyshire, DE4 3NA. (01629 580381)

PONTYPOOL & BLAENAVON RAILWAY.
The Council Offices, High Street, Blaenavon, Gwent, NP4 6PT. (01495 792263)

PLYM VALLEY RAILWAY.
Marsh Mills Station, Coypool Road, Plymouth, Devon, PL7 4NL.

RAILWAY AGE, THE.
Vernon Way, Crewe, Cheshire. (01270 212130)

ROGART STATION.
Rogart Station, near Golspie, Highland.

ROWDEN MILL STATION MUSEUM.
Private Site near Bromyard, Hereford and Worcester.

RUTLAND RAILWAY MUSEUM.
Cottesmore Iron Ore Mines Siding, Ashwell Road, Cottesmore, near Oakham, Leicestershire, LE15 7BX. (01572 813203)

SCOTTISH INDUSTRIAL RAILWAY CENTRE.
Minnivey Colliery, Dalmellington, Strathclyde. (01292 313579)

SEVERN VALLEY RAILWAY.
Railway Station, Bewdley, Hereford & Worcester, DY12 1BG. (01299 403816)

SOUTHALL RAILWAY CENTRE.
Glade Lane, Southall, Greater London, UB2 4PL. (0181 574 1529)

SOUTH DEVON RAILWAY.
Buckfastleigh Station, Buckfastleigh, Devon, TQ11 0DZ. (01364 642338)

SOUTHPORT RAILWAY CENTRE
Derby Road, Southport, Merseyside, PR9 0TY. (01704 530693)

SOUTH YORKSHIRE RAILWAY.
Barrow Road Railway Sidings, Barrow Road, Sheffield, S9 1LA.
(0114 242 4405/245 1214)

SPA VALLEY RAILWAY.
Eridge Station, Eridge Road, Eridge Green, Tunbridge Wells, Kent.

STEAMTOWN RAILWAY CENTRE.
Steamtown Railway Centre, Warton Road, Carnforth, LA5 9HX. (01524 732100)

STEAMTOWN, SCRANTON.
150 South Washington Avenue, Scranton, Pennsylvania, 18503, USA.

STOOMCENTRUM.
Maldegem, Belgium.

SWANAGE RAILWAY.
Station House, Railway Station, Swanage, Dorset, BH19 1HB. (01929 425800)

SWANSEA VALE RAILWAY.
Six Pit Junction, Nant-y-Ffin Road, Llamsamlet, Swansea, West Glamorgan.
(01222 613299)

STRATHSPEY RAILWAY.
Aviemore Speyside Station, Dalfaber Road, Aviemore, Highland, PH22 1PM.
(01479 810725)

SWINDON & CRICKLADE RAILWAY.
Tadpole Lane, Blunsdon, near Swindon, Wiltshire, SN2 4DZ. (01793 771615)

SWINDON GWR MUSEUM.
Farringdon Road, Swindon, Wiltshire. (01793 493189)

TELFORD HORSEHAY STEAM TRUST.
The Old Loco Shed, Horsehay, Telford, Shropshire, TF4 2LT. (01952 503880)

TRINIDAD GOVERNMENT RAILWAYS.
The system is currently disused.

VALE OF GLAMORGAN RAILWAY.
National Welsh Bus Garage, Broad Street, Barry, South Glamorgan.

VALE OF RHEIDOL RAILWAY.
Park Avenue, Aberystwyth, Dyfed, SY23 1PG. (01970 615993/625819)

WEST SOMERSET RAILWAY.
The Railway Station, Minehead, Somerset, TA24 5BG. (01643 704996)